Process Mining Techniques for Pattern Recognition

Process Mining Techniques for Pattern Recognition
Concepts, Theory, and Practice

Edited by

Vikash Yadav
Anil Kumar Dubey
Harivans Pratap Singh
Gaurav Dubey
Erma Suryani

CRC Press is an imprint of the
Taylor & Francis Group, an **informa** business

First edition published 2022
by CRC Press
6000 Broken Sound Parkway NW, Suite 300, Boca Raton, FL 33487–2742

and by CRC Press
2 Park Square, Milton Park, Abingdon, Oxon, OX14 4RN

© 2022 selection and editorial matter, Vikash Yadav, Anil Kumar Dubey, Harivans Pratap Singh, Gaurav Dubey, and Erma Suryani; individual chapters, the contributors

CRC Press is an imprint of Taylor & Francis Group, LLC

Reasonable efforts have been made to publish reliable data and information, but the author and publisher cannot assume responsibility for the validity of all materials or the consequences of their use. The authors and publishers have attempted to trace the copyright holders of all material reproduced in this publication and apologize to copyright holders if permission to publish in this form has not been obtained. If any copyright material has not been acknowledged please write and let us know so we may rectify in any future reprint.

Except as permitted under U.S. Copyright Law, no part of this book may be reprinted, reproduced, transmitted, or utilized in any form by any electronic, mechanical, or other means, now known or hereafter invented, including photocopying, microfilming, and recording, or in any information storage or retrieval system, without written permission from the publishers.

For permission to photocopy or use material electronically from this work, access www.copyright.com or contact the Copyright Clearance Center, Inc. (CCC), 222 Rosewood Drive, Danvers, MA 01923, 978–750–8400. For works that are not available on CCC please contact mpkbookspermissions@tandf.co.uk

Trademark notice: Product or corporate names may be trademarks or registered trademarks and are used only for identification and explanation without intent to infringe.

Library of Congress Cataloging-in-Publication Data
Names: Yadav, Vikash, editor.
Title: Process mining techniques for pattern recognition : concepts, theory, and practice / edited by Vikash Yadav, Anil Kumar Dubey, Gaurav Dubey, Harivans Pratap Singh, and Erma Suryani.
Description: First edition. | Boca Raton, FL : CRC Press, 2022. | Includes bibliographical references and index.
Identifiers: LCCN 2021043562 (print) | LCCN 2021043563 (ebook) | ISBN 9780367770495 (hbk) | ISBN 9780367770501 (pbk) | ISBN 9781003169550 (ebk)
Subjects: LCSH: Systems engineering. | Pattern recognition. | Data mining.
Classification: LCC TA168 .P68 2022 (print) | LCC TA168 (ebook) | DDC 620.001/171—dc23/eng/20211118
LC record available at https://lccn.loc.gov/2021043562
LC ebook record available at https://lccn.loc.gov/2021043563

ISBN: 978-0-367-77049-5 (hbk)
ISBN: 978-0-367-77050-1 (pbk)
ISBN: 978-1-003-16955-0 (ebk)

DOI: 10.1201/9781003169550

Typeset in Times
by Apex CoVantage, LLC

Contents

Preface vii
Editors ix
Contributors xi

Chapter 1 Concepts of Data Mining and Process Mining 1
Vineeta Singh & Vandana Dixit Kaushik

Chapter 2 Optimizing Web Page Ranks Using Query Independent
Indexing Algorithm 19
Shivi Panwar & Vimal Kumar

Chapter 3 Design and Implementation of Novel Techniques for
Content-Based Ranking of Web Documents 35
Ayushi Prakash, Sandeep Kumar Gupta, & Mukesh Rawat

Chapter 4 Web-Based Credit Card Allocation System Using Machine
Learning 47
Vipul Shahi, Yashasvi Srivastava, Utkarsh Sangam, Akarshit Rai, & Mala Saraswat

Chapter 5 Pattern Recognition 57
Akanksha Toshniwal

Chapter 6 Automated Pattern Analysis and Curated Sack Count
Leveraging Video Analysis on Moving Objects 69
Ritin Behl, Harsh Khatter, & Prabhat Singh

Chapter 7 DBSU: A New Fusion Algorithm for Clustering of Diabetic
Retinopathy Disease 83
Sanjay Kumar Dubey, Tanvi Anand, & Rekha Pal

Chapter 8 Dynamic Simulation Model to Increase the Use of Public
Transportation Using Transit-Oriented Development 97
Rizki Wahyunuari Ningrum, Erma Suryani, & Rully Agus Hendrawan

Chapter 9 Text Summarization Using Extractive Techniques 107
Mukesh Rawat, Mohd Hamzah Siddiqui, Mohd Anas Maan, Shashaank Dhiman, & Mohd Asad

Chapter 10 An Efficient Deep Neural Network with Adaptive Galactic
Swarm Optimization for Complex Image Text Extraction 121
Digvijay Pandey & Binay Kumar Pandey

Chapter 11 Diet Recommendation Model of Quality Nutrition for
Cardiovascular Patients 139
Surbhi Vijh & Sanjay Kumar Dubey

Chapter 12 Dynamic Simulation Model to Improve Travel Time
Efficiency (Case Study: Surabaya City) 153
*Shabrina Luthfiani Khanza, Erma Suryani, &
Rully Agus Hendrawan*

Index 165

Preface

Process mining primarily involves extracting process models from the event logs. The process in real life is more flexible and contains less structure. The older algorithms for process mining have issues in facing processes which are unstructured. The process models are spaghetti-like, which are quite difficult to comprehend. Construction of process models via raw traces, when done without any preprocessing, can be attributed as one of the reasons that lead to such an outcome. There are certain times when the system faces very similar behaviour or patterns while executing in an event log. The discovery of such process models can be improved by taking into consideration certain common patterns while invoking activities in traces. It can aid in explaining the concepts of relationships that exist between tasks. A lot of mining techniques are explained through the medium of this book, which will help organizations in unleashing and discovering their actual business processes. Of course, process mining goes beyond process discovery. It is plausible to find deviations, support decision-making, check conformance, predict delays, and recommend certain process redesigns by tightly coupling process models and data of an event. The text characterises and explors certain common model constructs in the event log. In order to give a definition and meaning to the results of these patterns, we adopt pattern definitions to capture these manifestations. This book provides an overview of the state of the art in process mining. It's mainly aimed to give an introduction to students, academics, and practitioners. This book, though meant for people who want to learn the basics and are new to these concepts, it also explains certain important concepts very rigorously. It is self-contained and covers the whole spectrum of process mining from process discovery to operational support. Those who have to deal with BI or BPM on an everyday basis can also refer to it. Because the techniques are very practical, pattern recognition and process mining can be put to use by utilizing the process mining software and event data in today's information systems. We hope that you find the book helpful and practically use the process mining practices which are available today.

Editors

Vikash Yadav received BTech degree from Uttar Pradesh Technical University, Lucknow, India, in 2009 and MTech degree from Motilal Nehru National Institute of Technology, Prayagraj, Allahabad, India, in 2013. He obtained a PhD degree from Dr. APJ Abdul Kalam Technical University, Lucknow, India, in 2017. Dr. Yadav is currently working as a Lecturer in the Department of Technical Education, Uttar Pradesh, India. He has published more than 50 articles in reputed SCIE and Scopus index journals. His research interests include image processing, data mining and machine learning. Dr. Yadav is a member of various technical research societies such as IEEE, ACM, CSI, IAENG, and so on.

Anil Kumar Dubey received a BTech degree from Uttar Pradesh Technical University Lucknow, India, in 2008 and MTech degree from Rajasthan Technical University Kota, India, in 2010. He obtained a PhD degree from the Career Point University Kota, India, in 2015. Dr. Dubey is an Assistant Professor (Selection Grade) in the CSE Department at ABES Engineering College, Ghaziabad, India. He has published several articles in reputed SCIE and Scopus index journals. His research interests include: human computer interaction, software engineering and artificial intelligence. Dr. Dubey is a member of various technical research societies, such as SMIEEE, MACM, IEEE CS, LMISTE, CSTA, and so on.

Harivans Pratap Singh received a BTech degree from Uttar Pradesh Technical University, Lucknow, India, in 2008 and an MTech degree from Uttarakhand Technical University, Dehradun, India, in 2013. He has a total of 12 years of experience in IT industry and teaching altogether. He has worked as a corporate trainer in Tech Mahindra for software engineering and software testing (manual and automation). His research work includes latent fingerprint indexing, their segmentation and identification efficiency. Other areas of interest are software testing, RPA (robotics process automation) on UiPath Studio, UML et Design Patterns.

Editors

Gaurav Dubey has a total academic and research experience of 20 years; currently he is a Professor in the CSE Department at ABES Engineering College, Ghaziabad, U.P., since January 2018. He served as Assistant Professor (ASET CSE) at Amity University Noida campus for 15 years (2002–2017). He has published 6 patents and 33 research papers (02 SCI, 31 SCOPUS) in various international journals, conferences and book chapters. He completed his PhD in 2017 from Amity University, Noida, in CSE domain (research specialization machine learning). He has various professional memberships like IEEE, ACM, IETE.

Erma Suryani received her BSc and MT degrees from the Institut Teknologi Sepuluh Nopember (ITS), Indonesia, in 1994 and 2001, respectively. She obtained her PhD degree from the National Taiwan University of Science and Technology (NTUST), Taiwan, in 2010. She has been working as a Professor at the Institut Teknologi Sepuluh Nopember (ITS), Indonesia, since 2005. Her current research interests are model-driven decision support systems, supply chain management, operation research, system dynamics with their applications in food security and sustainable transport.

Contributors

Tanvi Anand
Amity School of Engineering and Technology
Amity University
Noida, Uttar Pradesh, India
Email: tanvi29anand@gmail.com

Mohd Asad
Meerut Institute of Engineering & Technology
Meerut, U.P., India
Email: mohd.asad.cse.2017@miet.ac.in

Ritin Behl
ABES Engineering College
Ghaziabad, India
Email: ritin.behl@abes.ac.in

Shashaank Dhiman
Meerut Institute of Engineering & Technology
Meerut, U.P., India
Email: shashaank.dhiman.cse.2017@miet.ac.in

Sanjay Kumar Dubey
Amity School of Engineering and Technology Amity University
Noida, Uttar Pradesh, India
Email: skdubey1@amity.edu

Sandeep Kumar Gupta
Dr. K.N. Modi University
Rajasthan, India
Email: sundeepkrgupta@yahoo.com

Rully Agus Hendrawan
Institut Teknologi Sepuluh Nopember, Indonesia
Email: ruhendrawan@gmail.com

Vandana Dixit Kaushik
Harcourt Butler Technical University
Kanpur, India
Email: vandanadixitk@yahoo.com

Shabrina Luthfiani Khanza
Institut Teknologi Sepuluh Nopember (ITS), Indonesia
Email: luthfianisk@gmail.com

Harsh Khatter
KIET Group of Institutions
Delhi NCR, Ghaziabad, India
Email: harsh.khatter@kiet.edu

Vimal Kumar
Meerut Institute of Engineering & Technology
Meerut, U.P., India
Email: vimal.kumar@miet.ac.in

Mohd Anas Maan
Meerut Institute of Engineering & Technology
Meerut, U.P., India
Email: mohd.anas.cse.2017@miet.ac.in

Rizki Wahyunuari Ningrum
Institut Teknologi Sepuluh Nopember, Indonesia
Email: rizkiwahyunuari@gmail.com

Rekha Pal
Amity School of Engineering and Technology
Amity University
Noida, Uttar Pradesh, India
Email: palrekha106@gmail.com

Binay Kumar Pandey
G.B. Pant University of Agriculture and Technology
Pantnagar, Uttrakhand, India
Email: binaydece@gmail.com

Digvijay Pandey
Dr. A.P.J. Abdul Kalam Technical University
Lucknow, Uttar, India
Email: digit11011989@gmail.com

Shivi Panwar
Meerut Institute of Engineering & Technology
Meerut, U.P., India
Email: sp.shivipanwar@gmail.com

Ayushi Prakash
Dr. K.N. Modi University
Rajasthan, India
Email: ayushi.prakash@miet.ac.in

Akarshit Rai
ABES Engineering College
Ghaziabad, India
Email: akarshit.18bcs3002@abes.ac.in

Mukesh Rawat
Meerut Institute of Engineering & Technology
Meerut, U.P., India
Email: mukesh.rawat@miet.ac.in

Utkarsh Sangam
ABES Engineering College
Ghaziabad, India
Email: utkarsh.17bcs1171@abes.ac.in

Mala Saraswat
ABES Engineering College
Ghaziabad
Email: mala.saraswat@abes.ac.in

Vipul Shahi
ABES Engineering College
Ghaziabad, India
Email: vipul.17bcs1157@abes.ac.in

Mohd Hamzah Siddiqui
Meerut Institute of Engineering & Technology
Meerut, U.P., India
Email: mohd.hamzah.cse.2017@miet.ac.in

Prabhat Singh
ABES Engineering College
Ghaziabad, India
Email: prabhat.singh@abes.ac.in

Vineeta Singh
Harcourt Butler Technical University
Kanpur, India
Email: cs.vineeta.singh@gmail.com

Yashasvi Srivastava
ABES Engineering College
Ghaziabad, India
Email: yashasvi.17bcs1084@abes.ac.in

Erma Suryani
Institut Teknologi Sepuluh Nopember, Indonesia
Email: erma.suryani@gmail.com

Akanksha Toshniwal
PES University
Bangalore, India
Email: akanshatoshniwaleca@gmail.com

Surbhi Vijh
KIET Group of Institutions
Ghaziabad, Uttar Pradesh, India
Email: surbhivijh428@gmail.com

1 Concepts of Data Mining and Process Mining

Vineeta Singh & Vandana Dixit Kaushik

CONTENTS

1. Data Mining and Process Mining: An Overview 1
   1.1 Introduction 1
2. Related Work 2
   2.1 Data Mining Schemes 2
   2.2 Basic Rules and Techniques for Process Mining 3
3. Data and Issues in Educational Data Mining 5
4. Strategies That Rely on Data Mining and Process Mining in e-Learning Systems and EDM 8
5. Data Mining and Process Mining Strategies–Based Software Products 11
6. Some EDM Conferences and Journals 12
7. Conclusion 14
References 15

1 DATA MINING AND PROCESS MINING: AN OVERVIEW

1.1 INTRODUCTION

In many areas, there is a large amount of data about various processes involved in many modern information systems. For example, if we consider an e-learning system, a modern information system, it also accumulates as well as stores event data happening in the event logs. With the utilization of schemes such as data mining as well as process mining, event log data is consumed for performing better as well as analyzing the processes. Data gained via real processes is tested better by advanced software utilized for data mining as well as process mining. Due to immense increment in the amount of recorded data and event data in information systems, a keen interest in data mining as well as process mining has been developed. Here event data elucidates about information in detail processes history. Urge of improvement and to support business processes is also a root cause of developing keen interest in data mining as well as process mining, since improvement and enhancement of business processes in a dynamically changing scenario, which is competitive as well, is not easy. Data mining and process mining are complementary schemes to each other. Identified process models as well as aligned along with event log data assures the data analysis value and further yields a support for developing data mining and process mining schemes.

DOI: 10.1201/9781003169550-1

2 RELATED WORK

Data is the main essence in data mining and process mining. Both utilize many common mathematical techniques as well as schemes. Data mining mainly functions with the data, whereas process mining utilizes event data, including information for processes [1].

2.1 Data Mining Schemes

A data mining scheme with a multidisciplinary approach has evolved out of the many scientific fields, for example, artificial intelligence, machine learning, applied statistics, database theory, pattern recognition, algorithms and so on (see Figure 1.1). Data mining steps (see Figure 1.2) may comprise following:

- To identify associations as well as patterns, i.e. free search.
- Utilization of association schemes for guessing unknown values, i.e. predictive analytics.
- Detecting as well analyzing the exceptions associated with identified schemes i.e. detection of anomaly.

Definition provided by the Gartner Group:

> The method of identifying meaningful correlations, patterns as well as trends via shifting through huge amounts of data stored in repositories. Data mining involves pattern recognition technologies, along with statistical as well as mathematical schemes [2].

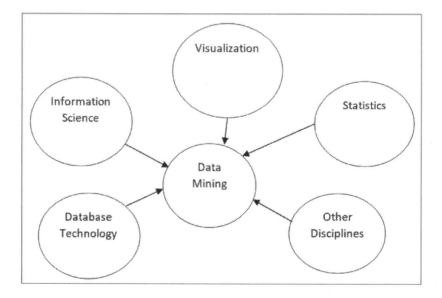

FIGURE 1.1 Data mining strategies along with allied scientific fields.

Concepts of Data Mining and Process Mining

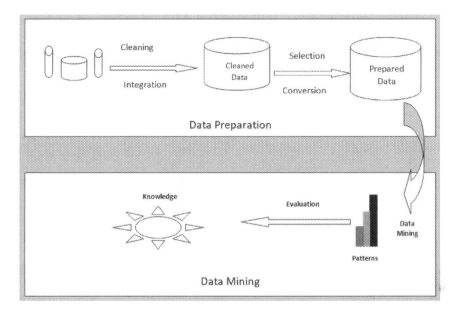

FIGURE 1.2 Data mining steps.

Definition provided by SAS Institute:

> Data mining is a methodology to discover anomalies, patterns as well as correlations among huge datasets for estimating outcomes/results. With a utilization of many schemes, this information may be further utilized to magnify revenues, cost cutting, for improvement of customer relationships, to minimize risks/dangers and so on [3].

Data mining may be considered as the computational methodology to identify patterns in huge data sets by including techniques of machine learning, statistics and artificial intelligence as well as database systems [4]. Data mining techniques as well as algorithms are composed of Bayesian networks, cluster analysis, nearest neighbor strategy, artificial neural networks, i.e. ANN, data visualization strategies, genetic algorithms as well as evolutionary programming, decision trees, support vector machine, decision trees, regression analysis, symbolic rules, linear regression and so on. Mainly mathematical models are the main constituent in analysis with the data mining concept. There is a possibility to utilize these strategies further to resolve different concrete issues as a result of suitable presence of software as well as hardware.

2.2 Basic Rules and Techniques for Process Mining

Research in process mining is in its early stage at the present time. The basic theme behind the concept of process mining is identification, controlling as well as improvement in real executing processes present in the advanced information systems via extraction of meaningful information taken from logs of events [5]. Big data as well as data mining involves a space in between that's filled by process mining, while at another place, it is situated in between business process modeling as well as analysis. There is a great opportunity for further theoretical as well as practical research and development in this field via the presence of huge data volume generated by business as well as business logic deployment at the each business level. Data science principles involvement at many aspects in the business processes depicts a novel way for management as well as modeling. With the help of information systems, a large amount of data for business processes is recorded in the name of event records, i.e. event logs, which can be further utilized in the form of source information in the context of retrieval of business process methodologies. Most of the time, event data present in the various organizations suffers from the issue of absence of their real-life process understanding. There is a possibility of conversion of hidden information about event logs to useful knowledge.

Process mining involves identification of automated processes, i.e. to extract process models out of event logs, checking of conformance, i.e. to monitor deviations via comparison of models as well as logs of events, describing the structure of organization, simulation models building in an automated manner, extension of models as well as retrieval, to forecast the behavior of a process so that a suggestion list may be built on the behalf of history of processes (see Figure 1.3). This technology is versatile in nature and can be incorporated upon any kind of operation process in various systems as well as organizations, although it is a recently developed technique. Process mining methodologies yield different ways to identify and supervise as well as improve the processes in different application areas; apart from this, it also provides ways to strict conformance evaluation, testing and credibility of information

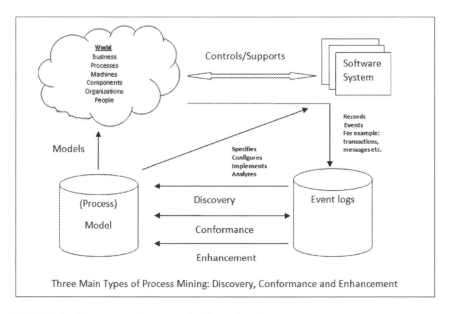

FIGURE 1.3 Process mining, a sample illustration diagram.

Concepts of Data Mining and Process Mining

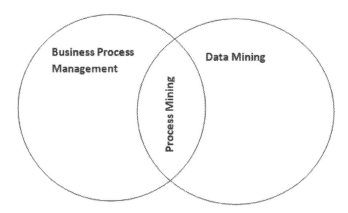

FIGURE 1.4 Data mining and process mining.

for fundamental organization processes. There are two points of concern in the context of any organization in the modern era: at one side, event data is growing tremendously; at another side, there should be proper alignment as per requirement of the perfect customer service. Thus this becomes a very pertinent methodology for current-time modernized organizations to handle the management of non-trivial functionary processes (see Figure 1.4).

Educational data mining, abbreviated as EDM, is in the essential application area in the field of data mining methodologies. Here the basic aim of the educational data mining concept is the utilization of large amounts data for educational processes, generated via various sources with varying formats at different stages of detail. The utilization of data is done in representation of information for different educational processes as well as for creating a more evident grasp of learning methodology and further enhancing its result outcomes.

3 DATA AND ISSUES IN EDUCATIONAL DATA MINING

There are various variations in the information systems as well environments in EDM pertaining to the issues and data (see Figure 1.7). Computer-based education involves utilization of computers in the field of education for delivering controlling instructions as well as education to the student. Computer-based education models/systems function on a stand-alone basis, i.e. these computer-based education models function on the computer without utilizing artificial intelligence in adaptation or student modeling as well as personalization and many more (see Figure 1.5). The internet is utilized globally, as a result of which so many educational system models have been devised that rely on the web, for example, online training models, distance learning models, e-learning models etc. On the other hand, development of artificial intelligence–based techniques and advancement has facilitated the evolution of smart/intelligent as well as adaptive educational models. Some of them are named adaptive intelligent hypermedia systems or AIHS, learning management systems or LMS, test/examination systems, intelligent tutoring systems or ITS as well

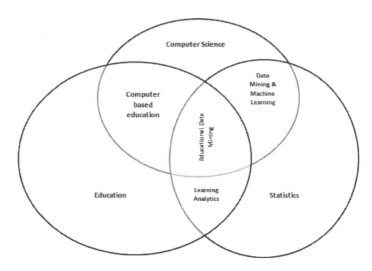

FIGURE 1.5 Educational data mining (EDM): Illustration diagram.

as quiz systems, hypermedia systems and many more [6–10]. Every model out of them delivers various data sources to be processed in a variety of ways that rely on the existing data type as well as special issues and tasks which need to be resolved via data mining strategies. Consumption of educational model data takes place by the researchers in educational data mining, for example, training/learning fulfilled with the help of intelligent computer systems, online classes, school information systems, electronic manuals, distance learning systems, computer-based testing systems, online tutorials, online educational forums, online educational modules, discussion forums, etc. [11].

Data possess special features/characteristics like hierarchy at several levels, for example at the level of grade system, at the level for query and at the level for subject; next is context, i.e. about a specific scenario like a particular student responds to a particular query at a particular date and time; next is duration data, i.e. recording of data for various resolutions to perform different analysis, such as recording of data at an interval of 20 seconds; the next one is long-duration-of-time data, i.e. a large amount data to be recorded for various sessions for an extended duration such as to cover year-long courses as well as semester-duration courses [12].

Analysis with the help of EDM is accompanied via any kind of information system abbreviated as IS. IS may be there to facilitate learning/training or education, for example, schools, universities, colleges, academic institutions, training institutions, professional education institutions, delivering training sessions via traditional or advanced/modern methodologies/techniques, informal training and so on. All this data is not only accompanied at individual-level student interactions to the educational institutions (data entry during tests, monitoring/navigating over the training/learning as well as testing modules, exercises in an interactive way) but also involves data related to coordination/cooperation among the students, for example, text chatting; data related to administrative tasks, i.e.

teacher, trainer, school, college, district and so on; demographics like school classes/lectures, age and gender etc.; emotionality of the student, for example, emotional state, motivation/encouragement etc. and many more. Data mining is mainly utilized to enhance the learning/training quality in the educational field. In the educational field, determination of quantitative measuring parameters is difficult compared to other fields. Outcomes need to be assessed/evaluated/ tested via performance pointers like improving/enhancing efficiency. In order to enhance and improve the current lecture materials and training content delivery, educational model decisions based on data are framed/formed. Utilization of EDM is accompanied in educational models and educational programs to resolve the issues in student behavior modeling as well as in forecasting of outcomes/ results of courses.

Various issues may be resolved through utilization of EDM. Examples include:

- Monitoring of learning progress in identifying unwanted behavior shown by students/learners on real-time basis, for example, the training termination, misuse of educational forums, fraud, lack of motivation, abuse and many others; showing warnings to the concerned parties; feedback process system provision to the trainers/teachers so that decision-making may be improved for student training/learning and to adapt preemptive steps to remedy the arisen issue [13].
- To forecast/identify achievement of student, evaluation/testing of learning/ training as well as knowledge acquired, to draft suggestions for the students on the basis of interests as well as activities of students/learners during the learning/training process [14].
- Approaches at individual level, i.e. adaptation of training for every student/ learner involving course material/content, structural navigation over the course, content/material presentation methodology, detecting student groups as per their individual-level features/characteristics, characteristics at personal level, training features/advantages and so on [15–18].
- Creating educational content structure as well as curriculum, future courses scheduling as well as planning, planning of course, resource allocation planning, creating structures for the way of accessing learning content, curriculum creation/development, planning of consultations etc. [19–21].
- Devising as well as validating theories of science based on learning techniques, formulating novel scientific hypotheses, enhancement/advancement of domain lecture delivery style such as skills, expertise, concepts, modules designed for training as well as their correlation [22,23]. Student cognitive models demonstrating skills as well knowledge, estimating parameters for probability-based models relying on learning data to estimate/identify the possibility of events of interest [24,25]. The varying nature of issues as well as their performance on education reaches towards the requirement of adapting data mining along with process mining strategies/techniques for these issues and data resolution. In the education area, data mining techniques applicability has been demonstrated [26].

4 STRATEGIES THAT RELY ON DATA MINING AND PROCESS MINING IN E-LEARNING SYSTEMS AND EDM

In educational data mining, the techniques mostly involve classification, clustering, text mining, i.e. text analytics as well as data mining, relationship mining, Bayesian theory model, anomaly detection, analysis of social networks, knowledge tracing, discovery along with models, data distillation for human decision/judgment, non-negative matrix factorization, process mining strategies i.e. probabilistic techniques, alpha algorithms, genetic algorithms, heuristic algorithms and so on. Here prediction refers to the type of dependence of a target attribute over combination of other attributes. A different kind of prediction strategies involve a classification strategy, i.e. category refers to target variable, regression, i.e. background variables, as well as target variables that are numbers, or density score, i.e. value of prediction is known as probability density function. Utilization of these strategies for predicting performance of students and determining student behavior patterns is illustrated in Figure 1.6. [27,28]

Clustering refers to the detection of similar instance groups. For determination of similarity, utilization of distance measure is done. After the determination of cluster sets, classification of new items is accompanied as per the nearest cluster. In case of EDM, the clustering is done for grouping similar course contents/materials as well

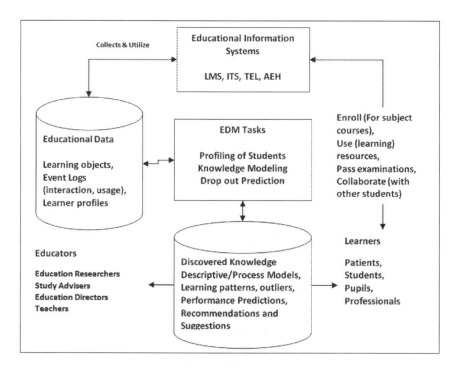

FIGURE 1.6 EDM and strategies that rely on data mining/process mining in an e-learning system.

Concepts of Data Mining and Process Mining

as in formation of student groups as per the pattern of interaction of students or as per the student knowledge [29,30]. Different kinds of clustering strategies with their applications are demonstrated in [31]. High-quality information production from the text is done with the help of text mining strategy. Specific tasks involve text clustering, extraction of entity/concept, sentiment analysis, text mining classification of text, summarization of document, generation of granular taxonomies as well as with the help of entity relationship model. Utilization of text mining strategy is used for analyzing discussion board content, chats, forums, documents and web pages etc. in EDM.

The relationship in between variables is established with the help of relationship mining, and it also facilitates their presentation in the form of rules for further utilization. Different kinds of relationship mining are present. Some of them are sequential pattern mining, i.e. association on a temporary basis in between variables; association rule mining, i.e. variable relations in between; correlation mining, i.e. correlation on a linear basis in between variables; and casual data mining, i.e. relationship between variables on casual basis. The role of relationship mining involves determination of student behavior relationships, i.e. behavioral patterns, in diagnosis of teaching difficulties as well as in diagnosing mistakes occurring simultaneously [32, 33].

For evaluating/testing of student skills, there is a popular technique known as knowledge tracing, i.e. KT, utilized effectively in cognitive tutor systems [34]. Utilization of KT is accompanied via a cognitive-based model which uses mapping techniques for issue-resolving items that need skills/expertise and further records right as well as wrong answers/replies of students as a proof of student knowledge on a specific skill. For some time, it monitors knowledge of students, and further conversion is done into four variables.

KT resembles the Bayesian theory technique.

Social network analysis, i.e. SNA, is utilized for understanding and measuring relationships in between entities in the network information. Social network analysis

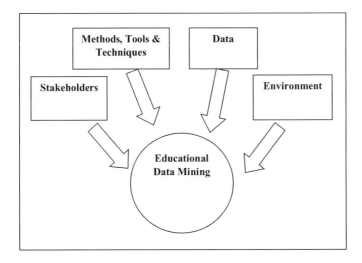

FIGURE 1.7 Educational data mining: A scenario.

contemplates social relationships in the form of networking theory comprising nodes, i.e. representation of actors on individual basis inside the network; ties or connections, i.e. representation of relationships between individuals like organizational post, friendship and so on. In the case of EDM, the utilization of SNA goes towards gaining information for interpreting as well as analyzing of structure and relationship during the task of interaction as well as during communicative interactions [35].

Identification/detection of outlier – Identification of data which is significantly different from the other remaining data. Abnormal values resemble the observations/measurements, often less or more as compared to other values. Students who have learning issues, variations in the behavioral patterns/actions of the teachers as well as students and identification of learning processes on an irregular basis may be detected via the technique of EDM anomaly detection/identification [36]. It is a searching/discovery process along with models to utilize an already-examined phenomena model via clustering, prediction or knowledge engineering on a manual basis as an element of other types of analysis, like relationship mining or prediction [37]. This technique has been utilized most of the time in educational data mining, further facilitating detection of relationships between behavioral patterns of students and characteristics of them, in machine learning strategies, the utilization of psychometric modeling techniques, and research analysis in different fields of study/research [38]. For facilitating decision-making processes as well as in extraction of utilitarian information via techniques such as visualization, generalization as well as interactive interfaces for demonstrating data in an easily understandable form comes under the name of distillation of data for human verdict/judgment. This strategy involves generating statistical data for the learning process in determination of basic features/characteristics and generating summarized reports as well as data on a trainee's behavioral pattern. Strategies like graphical schemes as well as data visualization help us for understanding, exploring and seeing huge amount educational data on immediate basis. Data distillation for human decision-making has been utilized for assisting trainers/teachers via analyzing and visualizing activities of students as well as for information utilization in the EDM [39,40]. A technique utilized to demonstrate clearly via Q-matrix or transfer model is named the non-negative matrix factorization, i.e. NMF [41].

There exist various NMF algorithms for providing varying outcomes. Basically, NMF utilizes an array comprising of positive numbers, which is the multiplication generated via two matrices of smaller size, such as under observation, a learning process can involve a matrix representing the outcomes of student tests, further bifurcating into two matrices, namely A and B, where A may represent elements of learning, while B may represent skills for every student. Here, process mining exhibits its main agenda as full illustration of whole process along with its analysis as well as improvement via extracting knowledge for the process involved in learning models out of the event logs. Process mining may be utilized in representation of the behavioral pattern of students as per the records of the event log. For every event, there is data involving a time stamp as well as data involving a learning/training process. It may contain information for knowledge evaluation of students, information regarding participating in chats as well as forums, for lectures, for viewing other educational contents/materials, information for passing examinations/tests, data utilized to illustrate collaborative learning processes or information in regard to events involving meta cognitive prompts [42–45]. On the basis of behavioral patterns of students, they may be grouped into two sections. Here, it is essential to demonstrate the event concept (here, "event" may refer to a mouse click as well) as well as the series-of-events concept. Use of dotted chart diagrams is done most of the time in visualization of events on an individual basis. Moreover, a process model is created, and conformance checking/testing is done. In creation and testing of learning process models, the basic as well as specific process mining strategies have been adapted such as genetic algorithms, heuristic algorithms, probabilistic strategies and alpha algorithms. Moreover, data mining strategies also have been used. The demonstration of the process model is done often via BPMN model or Petri net. Complications occur while creating a learning process model due to the presence of parallel tasks, loops, "noise" presence and mutual effects of some of the tasks to other tasks. Although there is much research present in the field data mining, in Russian scientific journals, there are few papers relevant to the study of application of strategies that rely on data mining as well as process mining in the context of learning/training process. Some of them are utilization of an artificial neural network, i.e. ANN, for modeling educational process in high school; use of cluster analysis strategy in studying structure/organization of high school students/learners values; utilization of techniques pertaining to educational data mining as well as learning analytics for educational qualifications; and use of analysis of variance strategy was done in studying adaptation factors of students to training situations/conditions [46–49].

A summarized overall demonstration for the strategies and techniques pertaining to data mining in education field as well as utilization of classification strategies in data analysis for training models has been demonstrated in [50].

5 DATA MINING AND PROCESS MINING STRATEGIES–BASED SOFTWARE PRODUCTS

There is a requirement of specific software in implementation of data mining as well as process mining strategies [51]. Such characteristics are added by the various software vendors in their software. Table 1.1 illustrates such list of software products.

ProM is basic freeware software available with 1,500 plug-ins and further facilitates utilization of various techniques as well as strategies for data mining and process mining, various kinds of models, data for conversion of models and data etc. The ProM Lite version exhibits mostly basic and used modules. Normally, software products available commercially are easy to use, or user friendly. Around 40 software tools most utilized in the educational field are described in the research paper [52].

TABLE 1.1
Software tools possessing features of data mining and process mining

S. No.	URL	Availability/Vendor Name	Software Name
1	orange.biolab.si	Open Source	Orange
2	www-01.ibm.com/software	IBM	SPSS
3	www.celonis.de	Celonis GmbH	Celonis Process Mining
4	www.cs.waikato.ac.nz/ml/weka/	Open Source	WEKA
5	www.exeura.eu	Exeura	Rialto Process
6	www.fluxicon.com	Fluxicon	Disco
7	www.lexmark.com	Lexmark	Perceptive Process Mining
8	www.minitlabs.com	Gradient ECM	Minit
9	www.nltk.org	Open Source	NLTK
10	www.promtools.org	Open Source	ProM
11	www.promtools.org	Open Source	ProM Lite
12	www.qpr.com	QPR	QPR ProcessAnalyzer
13	www.rapidprom.org	Open Source	RapidProM
14	www.rapidminer.com	Open Source	RapidMiner
15	www.snp-bpa.com	SNP AG	SNP Business Process Analysis

TABLE 1.2
Scientific conferences based on EDM, starting from first conference

S. No.	Conference	Organizing Location
1	EDM2008 (First conference)	Montreal, Canada
2	EDM2009	Cordoba, Spain
3	EDM2010	Pittsburgh, United States
4	EDM2011	Eindhoven, Netherlands
5	EDM2012	Chania, Greece
6	EDM2013	Memphis, United States
7	EDM2014	London, United Kingdom
8	EDM2015	Madrid, Spain
9	EDM2016	Raleigh, United States
10	EDM2017	Wuhan, China

6 SOME EDM CONFERENCES AND JOURNALS

EDM has emerged as stand-alone area of research. It involves research for training intellectual models, i.e. artificial intelligence, in education, user modeling, intelligent tutoring systems, adaptive and intelligent educational hypermedia as well as technology-enhanced learning. Table 1.2 and Table 1.3 demonstrate summarized lists of some conferences, while Table 1.4 lists some journals in the area of educational data mining.

TABLE 1.3
List of some conferences in the field of EDM

S. No.	Type	Organizing Year	Conference Name
1	Annual basis	2011	Conference Title "International Conference on Learning Analytics and Knowledge"
2	Every two years	1983	Conference Title "International Conference on Artificial Intelligence in Education"
3	Annual basis	2009	Conference Title "International Conference on User Modeling, Adaptation, and Personalization"
4	Every two years	1988	Conference Title "International Conference on Intelligent Tutoring Systems"
5	Annual basis	2008	Conference Title "International Conference on Educational Data Mining"

TABLE 1.4
List of some journals in the field of EDM

S. No.	Publisher	Journal Name
1	ACM	*ACM Special Interest Group on Knowledge Discovery and Data Mining, Explorations*
2	Springer	*Data Mining and Knowledge Discovery*
3	AIED Society	*International Journal of Artificial Intelligence in Education*
4	IGI Global	*International Journal of Data Warehousing and Mining*
5	IEEE	*IEEE Transactions on Knowledge and Data Engineering*
6	Wiley	*Statistical Analysis and Data Mining*
7	Elsevier	*Internet and Higher Education*
8	SAGE Publications	*American Educational Research Journal*
9	Taylor & Francis	*Journal of the Learning Sciences*
10	Australasian SOC	*Australasian Journal of Educational Technology*
11	Elsevier	*Computer and Education*
12	EDM Society	*Journal of Educational Data Mining*
13	Wiley	*Wiley Interdisciplinary Reviews—Data Mining and Knowledge Discovery*
14	IEEE	*IEEE Transactions on Learning Technologies*
15	SAGE Publications	*Journal of Educational and Behavioral Statistics*
16	POLSKA	*Archives if Mining Sciences*
17	IEEE	*IEEE Transactions on Education*
18	Springer	*User Modeling and User-Adapted Interactions*

7 CONCLUSION

In this chapter, the authors have described fundamental principles of research in the context of the EDM field. Here authors have listed many examples to demonstrate the issues resolved through data mining and process mining strategies in conventional learning as well as e-learning. Different methods along with their possibilities and constraints have been elucidated. A brief summary of educational data mining strategies–based major scientific journals and conferences have been illustrated. The concept of educational data mining helps to navigate/investigate learning content in e-learning system modules as well as the process executed in it.

In the education field, with the utilization of information and communication technologies, there is the generation of huge amounts of data involving comprehensive information related to students as well as processes via which they enter into educational courses. To learn the understanding capability of students, data consumed and derived via teachers/instructors and factors influencing the performance and skills of the students may be tested/examined. In search of solutions for all such issues, researchers are exploring data mining strategies in education. Educational data mining is focused on developing particular techniques to have a study of educational databases emerging via any kind of information system facilitating education or training (i.e. traditional training/education, informal learning/teaching or modern/advanced training/education delivering traditional as well as modern teaching/training methodologies in vocational training institutions, colleges, universities, schools and so on).

With the help of EDM, there has been a close collaboration/relationship between researchers and practitioners from the various fields such as computer science, psychology, statistics, education and psychometrics. The major involvement of process mining is in identifying, observing and enhancing real processes via knowledge extraction out of the event logs automatically as per the information system record. Educational issues may be addressed through this strategy.

The basic aims are:

From huge amount of education event logs, knowledge relevant to processes is extracted, for example, a bunch of curriculum pattern templates or models related to key performance indicators.

Requirements and educators' hypotheses, analyzing processes of education to test to what extent they resemble the standard curriculum limitations.

Improving educational process strategies with key performing pointers/indicators such as decision point, time of execution, bottlenecks and so on.

Educational processes may be personalized as per the guidelines of the best course unit recommendation or via learning paths for students/learners as per learner profile, as per priorities of the learner or as per the skill/expertise of the learner as well as online identification of negligence/violation as per standard requirements/prerequisites. Further, it can be summarized that the complementary utilization of data mining and process mining strategies in e-learning may further enhance/improve the teaching quality and lecture delivering quality and may enhance its access, reach, availability and effectiveness.

REFERENCES

[1] Janssenswillen G, Depaire B, Swennen M, Jans M, Vanhoof K. bupaR: Enabling reproducible business process analysis. *Knowledge-Based Systems.* 2019 Jan 1;163:927–930.

[2] Gartner Inc. IT Glossary. URL: www.gartner.com/it-glossary/data-mining (21.01.2017).

[3] SAS Institute Inc. URL: www.sas.com/en_us/insights/analytics/data-mining.html (21.01.2017).

[4] Romero C, Ventura S. Educational data mining and learning analytics: An updated survey. *Wiley Interdisciplinary Reviews: Data Mining and Knowledge Discovery.* 2020 May;10(3):e1355.

[5] Cerezo R, Bogarín A, Esteban M, Romero C. Process mining for self-regulated learning assessment in e-learning. *Journal of Computing in Higher Education.* 2020 Apr;32(1):74–88.

[6] Mahajan G, Saini B. Educational data mining: A state-of-the-art survey on tools and techniques used in EDM. *International Journal of Computer Applications & Information Technology.* 2020;12(1):310–316.

[7] Hung HC, Liu IF, Liang CT, Su YS. Applying educational data mining to explore students' learning patterns in the flipped learning approach for coding education. *Symmetry.* 2020 Feb;12(2):213.

[8] Mostow J, Beck J. Some useful tactics to modify, map and mine data from intelligent tutors. *Journal of Natural Language Processing.* 2006;12:195–208.

[9] Merceron A, Yacef K. Mining student data captured from a web-based tutoring tool: Initial exploration and results. *Journal Of Interactive Learning Research.* 2004;15:319–346.

[10] Fister Jr I, Fister I. Information cartography in association rule mining. arXiv preprint arXiv:2003.00348. 2020 Feb 29.

[11] Salloum SA, Alshurideh M, ElnAgar A, Shaalan K. Mining in educational data: Review and future directions. In Joint European-US Workshop on Applications of Invariance in Computer Vision. Springer, Cham, 2020 Apr 8; 92–102.

[12] Romero C, Ventura S. Data mining in education. *The Wiley Interdisciplinary Reviews: Data Mining and Knowledge Discovery.* 2013;3:12–27.

[13] Kotsiantis S, Patriarcheas K, Xenos MN. A combinational incremental ensemble of classifiers as a technique for predicting student's performance in distance education. *Knowledge-Based Systems.* 2010;23:529–535.

[14] Tang TY, Daniel BK, Romero C. Recommender systems in social and online learning environments. *Expert Systems.* 2015;32(2):261–263.

[15] Romero C, Ventura S. Preface to the special issue on data mining for personalised educational systems. *User Model User-Adapted Interact.* 2011;21:1–3.

[16] Bannert M, Reimann P, Sonnenberg C. Process mining techniques for analyzing patterns and strategies in students' self-regulated learning. *Metacognition and Learning.* 2014;9(2):161–185.

[17] Bouchet F, Harley JM, Trevors GJ, Azevedo R. Clustering and profiling students according to their interactions with an intelligent tutoring system fostering self-regulated learning. *Journal of Educational Data Mining.* 2013;5(1):104–146.

[18] Ayers E, Nugent R, Dean N. A comparison of student skill knowledge estimates. International Conference on Educational Data Mining. Cordoba, Spain, 2009; 1–10.

[19] Cairns AH, Gueni B, FhiMa M, CAirns A, David S, Khelifa N. Towards custom-designed professional training contents and curriculums through educational process Mining. The Fourth International Conference on Advances in Information Mining and Management, 2014; 53–58. Copyright (c) IARIA, 2014. ISBN: 978-1-61208-364-3.

[20] Garcia E, Romero C, Ventura S, Castro C. Collaborative data mining tool for education. International Conference on Educational Data Mining. Cordoba, Spain, 2009; 299–306.
[21] Hsia T, Shie A, Chen L. Course planning of extension education to meet market demand by using datamining techniques: An example of Chinkuo Technology University in Taiwan. *Expert Systems with Applications.* 2008;34:596–602.
[22] Siemens G, Baker RSJD. Learning analytics and educational data mining: Towards communication and collaboration. Proceedings of the 2nd International Conference on Learning Analytics and Knowledge. Vancouver, British Columbia, Canada, 2012; 1–3.
[23] Pavlik P, Cen H, Koedinger K. Learning factors transfer analysis: Using learning curve analysis to automatically generate domain models. Int Conf Edu Data Min. 2009; 121–130.
[24] Frias-Martinez E, Chen S, Liu X. Survey of datamining approaches to user modeling for adaptive hypermedia. *IEEE Transactions on Systems, Man, and Cybernetics, Part C (Applications and Reviews).* 2006;36(6):734–749.
[25] Wauters K, Desmet P, Noortgate W. Acquiring item difficulty estimates: A collaborative effort of data and judgment. International Conference on Educational Data Mining. Eindhoven, The Netherlands, 2011; 121–128.
[26] Baker RS, Inventado PS. Educational data mining and learning analytics. *Learning Analytics.* New York, NY: Springer, 2014;61–75.
[27] Romero C, Espejo PG, Zafra A, Romero JR, Ventura S. Web usage mining for predicting final marks of students that use Moodle courses. *Computer Applications in Engineering Education.* 2013;21(1):135–146.
[28] Baker RSJD, Gowda SM, Corbett AT. Automatically detecting a student's preparation for future learning: Help use is key. Fourth International Conference on Educational Data Mining. Eindhoven, The Netherlands, 2011; 179–188.
[29] Bogarín A, Romero C, CeRezo R, Sánchez-Santillán M. Clustering for improving educational process mining. Proceedings of the Fourth International Conference on Learning Analytics and Knowledge. ACM, New York, NY, USA, 2014; 11–15.
[30] Vellido A, Castro F, Nebot A. Clustering educational data. *Handbook of Educational Data Mining.* Boca Raton, FL: Chapman and Hall/CRC Press, 2011; 75–92.
[31] Dutt A, Aghabozrgi S, Ismail MAB, Mahroeian H. Clustering algorithms applied in educational data mining. *International Journal of Information and Electronics Engineering* 2015;5(2):112–116.
[32] Tane J, Schmitz C, Stumme G. Semantic resource management for the web: An e-learning application. International Conference of the WWW. New York, 2004; 1–10.
[33] Merceron A, Yacef K. Measuring correlation of strong symmetric association rules in educational data. *Handbook of Educational Data Mining.* Boca Raton, FL: CRC Press, 2011; 245–256.
[34] Corbett A, Anderson J. Knowledge tracing: Modeling the acquisition of procedural knowledge. *User Model User-Adapted Interact.* 1995;4:253–278.
[35] Rabbany R, Takaffoli M, Zaïane O. Analyzing participation of students in online courses using social network analysis techniques. International Conference on Educational Data Mining. Eindhoven, The Netherlands, 2011; 21–30.
[36] Ueno M. Online outlier detection system for learning time data in e-learning and its evaluation. International Conference on Computers and Advanced Technology in Education. Beijing, China, 20041; 248–253.
[37] Charitopoulos A, Rangoussi M, Koulouriotis D. On the use of soft computing methods in educational data mining and learning analytics research: A review of years 2010–2018. *International Journal of Artificial Intelligence in Education.* 2020 Oct;30(3):371–430.

[38] Bakhshinategh B, Zaiane OR, ElAtia S, Ipperciel D. Educational data mining applications and tasks: A survey of the last 10 years. *Education and Information Technologies.* 2018 Jan;23(1):537–553.
[39] Baker RSJD. Data mining for education. *International Encyclopedia of Education*, 3rd Edition. Oxford, UK: Elsevier, 2010; 7: 112–118.
[40] Cantabella M, Martínez-España R, Ayuso B, Yáñez JA, Muñoz A. Analysis of student behavior in learning management systems through a Big Data framework. *Future Generation Computer Systems.* 2019 Jan 1;90:262–272.
[41] Desmarais MC. Mapping question items to skills with non-negative matrix factorization. *ACM SIGKDD Explor.* 2011;13:30–36.
[42] Trcka N, Pechenizkiy M, van der Aalst W. Process mining from educational data. *Handbook of Educational Data Mining.* Boca Raton, FL: CRC Press, 2011; 123–142.
[43] Alom BM, Courtney M. Educational data mining: A case study perspectives from primary to university education in Australia. *International Journal of Information Technology and Computer Science.* 2018;10(2):1–9.
[44] Schoor C, Bannert M. Exploring regulatory processes during a computer-supported collaborative learning task using process mining. *Computers in Human Behavior.* 2012;28:1321–1331.
[45] Sonnenberg C, Bannert M. Discovering the effects of metacognitive prompts on the sequential structure of SRL-processes using process mining techniques. *Journal of Learning Analytics.* 2015;2(1):72–100.
[46] Petrova MV, Anufrieva DA. Investigation of the possibilities of methods of intellectual data analysis in modeling the educational process in the university. *Vestnik Chuvashskogo Universiteta.* 2013;3:280–285. (in Russian).
[47] Avadehni YI, Kulikova OM, Radionova VA. The study of the structure of values of university students with the use of data mining technologies. *Sovremennye problemy nauki i obrazovaniya.* 2013;6:841. (in Russian).
[48] Veryaev AA, Tatarnikova GV. Educational data mining i learning analytics: Directions of development of educational qualification. *Prepodavatel' XXI vek.* 2016;2:150–160. (in Russian).
[49] Shumetov VG, Lyaskovskaya OV. Study of the factors of adaptation of the students of the 2000s to the training in the university by the methods of data Mining. *Srednerusskij vestnik obshchestvennyh nauk.* 2015;6:49–56. (in Russian).
[50] Gorlushkina NN, Kocyuba IY, Hlopotov MV. The tasks and methods of intellectual analysis of educational data to support decision-making. *Obrazovatel'nye tekhnologii i obshchestvo.* 2015;1:472–482. (in Russian).
[51] Calders T, Pechenizkiy M. Introduction to the special section on educational data mining. *Sigkdd Explorations.* 2012;13:3–6. doi:10.1145/2207243.2207245.
[52] Slater S, Joksimović S, KoVanovic V, Baker RSJD, Gasevic D. Tools for Educational data mining: A review. *Journal of Educational and Behavioral Statistics.* 2017;42(1):85–106.

2 Optimizing Web Page Ranks Using Query Independent Indexing Algorithm

Shivi Panwar & Vimal Kumar

CONTENTS

1 Introduction 19
    1.1 Organization of the Chapter 20
2 Related Work 20
3 Proposed Model 21
    3.1 Architecture 21
        3.1.1 Optimized Ranked Pages 25
4 Result Analysis 25
    4.1 Assumptions 25
    4.2 Input Web Graph 1 25
    4.3 Overall Result Analysis and Performance 31
5 Conclusion 32
References 32

1 INTRODUCTION

The information retrieval system [1] is a set of instructions combined as a computer program to extract information from the data warehouse that is present on the internet. When a user searches an 'x' item on the web, the search process starts, and the information retrieval system scans all the data on the web based on the item 'x' searched by the user. When 'x' is searched, it contains some specific keywords that play a crucial role in executing the retrieval of information from the web. Based on those keywords, the search engines provide a certain rank to the search result or, say, a web document that is considered in this paper [1]. The document with the first rank shows up on the top of the results, the second one comes next to the top, and so on and so forth. This assignment of rank to each document is done in two ways: query dependent rank assignment and query independent ranking [2].

    The challenge with query dependent [3] is that irrelevant results appear in the search, because when the search is based on keywords, it may bring results with a maximum number of keywords instead of relevance or authenticity of the document. However, with query independent search, the results are based on certain criteria including the number of pages linked to the resultant page/document, the relevancy of the content in the document, the authenticity of the document, and its usefulness depending on the number of visits [4,5].

In this chapter, we have discussed the query independent method [6] to rank the web documents while implementing a search. This discusses the query independent ranking method, which is based on the Markov Chain, and by this method, a rank is provided to each document in the collection of documents independent of query [7–9]. Therefore, it can also be computed offline. The following sections describe how we implemented the query independent ranking method. This algorithm is not meant to rank a website but to rank individually each web page of the website or web. This improves the results of each web page, and when the user searches for a specific document or content, the results show the most relevant web page instead of the entire website, which may or may not contain the relevant search result [10–12]. It is the recursive algorithm to calculate the rank of a web page, say A, defined by the web page's page ranks Ti, which links to Page A. The page rank of page T is always weighted by the number of outbound links C (T) on page T. That means more outbound links of page T less will be beneficial to page A. More web pages referring to page A will improve its page rank value. At the end, we have to add all page ranks contributing to page A and multiply it by a damping factor d, which can be set between 0 and 1. As a page rank algorithm uses link structure to count the rank value of the web page, some results produced by the page rank algorithm [13] are not relevant to the user's query; this problem is called theme drift [2].

1.1 Organization of the Chapter

The related work performed and executed by other scholars and scientists from around the world is present in Section 2 of this chapter. Section 3 contains our proposed algorithm, Efficient and Advanced Query Independent Page Ranking Algorithm, along with its working, implementation, and theoretical analysis to throw light on the work's importance. Section 4 depicts the practical analysis and application and the results of the algorithm in the real world. In the same section, multiple cases are considered to explore the variety in results while applying the algorithm to show the behavior of our proposed advanced algorithm under different scenarios. Section 5 of the chapter concludes with the results and analysis of the work done throughout the course of this chapter.

2 RELATED WORK

C.D. Manning et al. (2008) [1], gave a detailed introduction to the information retrieval system that drives the basics of the research carried out in this chapter. Sergey Brin and Lawrence Page (2012) [2] presented, for the first time, a query independent web structure–based ranking system. It has the lower relevancy theme drift problem, having a Moister Factor of 0.8, but exactly why 0.8 is selected was never disclosed nor shown in the calculations. H. Tirri et al. (2003) [14] produced the research that described the issues and challenges with the systems while carrying out web research in "Search

Optimizing Web Page Ranks

in Vain: Challenges for Internet Search." The work done by Kleinberg (1999) [15] focuses on hypertext induced topic search, which is a query dependent web content–based ranking system, but it has a theme drift problem. The analysis by Xing, Wenpu, et al. (2004) [6] showed a weighted page rank algorithm, which is a query independent and web structure mining–based ranking system. Kumar Gyanendra, Neelam Duhan, and A. K. Sharma (2011) [16] worked on page ranking based on the number of visits of links of web pages (VOL). It is a query independent, web structure, and web page visiting priority–based ranking system. It has a comparatively higher page relevance, but it comes with the theme drift problem. The scholars Simple Sharma et al. (2012) [17] presented an improved method for computation of page rank by reducing the computational complexity and theme drift problem. Y. Zhang, K. Nan, and Y. Ma (2008) [18] presented research on retrieving information from the web using ontology that considers the existence to become the reality. The importance of work done by Y. Lu et al. (2014) [19] is that they identified the importance of text features in the web content and extracted the same using genetic algorithms. R. Premalatha and S. Srinivasan (2014) [20] implemented text processing in an information retrieval system using the vector space model. The research worked on text extraction on the web pages to necessitate the page information rapidly.

3 PROPOSED MODEL

The proposed algorithm is a query independent ranking method based on the Markov chain [21,22]. By this method, a rank is provided to each document in the collection independent of query. Therefore, rank can be computed offline. And this method also helps in determining the quality of a page on the bases of authorization.

3.1 Architecture

When a user searches the web, he may get variable results, and these results may or may not be according to the requirement because of the malicious practices followed by the rankers to rank their website on the top [23]. So the proposed algorithm considers the results to be linked to a trusted source. If a website A, which is genuine and verified [24,25], refers to the search result, then it ranks better in the web according to this algorithm. The architecture is shown in Figure 2.1.

> **Web Search Input:** The user comes into the web and searches for a keyword to find the required web page. Let's say the search is carried out for 'x.' The web already has this data, and the search engines have to bring the optimal search results and rank them according to relevance. The pages on the web are present and hosted, and consider they are named Page 1, Page 2, Page i.
>
> **Search Execution:** Multiple pages from these contain the relevant information [24,25], while others don't. In an ideal state, the search engine brings up the results that are most relevant according to the search work 'x.' But what result will come in the first place depends on how the page has been ranked. Multiple sites rank their pages falsely to improve their clicks and visiting users. So it's necessary to see which one is the relevant result. In this algorithm, when the search is executed, the ranking algorithm

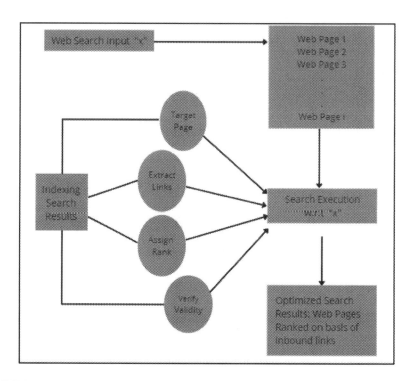

FIGURE 2.1 Architecture of proposed system.

ranks the pages Page 1, Page 2, Page 3, Page i according to the number of inbound links to them from verified sources.

Indexing Search Results—Proposed Algorithm: Web document rank of a record can be characterized as the small part of time that the surfer spent on that report, all things considered. The likelihood that the arbitrary surfer visits a report is its web document rank. Clearly, records that are hyperlinked by numerous different archives will be visited all the more frequently. Also, the likelihood that an irregular surfer will get exhausted and restart from some other arbitrary report is a dampness factor (say m) and with the likelihood of (1-m) follow the out connect picked arbitrarily.

Here N is the number of documents in the collection. "Prob" represent the probability transition matrix, and "adj" represents the adjacency matrix. Moist_fact is the Moister Factor, and x is the probability vector.

Initialize the Probability Transition Matrix

```
For i=0 to N do
  For j=0 to N do
  prob[i][j]=adj[i][j];
  end of loop
end of loop
```

Optimizing Web Page Ranks

Compute no of out links from a particular node

```
For i=0 to N do
        counter=0;
For j=0 to N do
                if(prob[i][j]==1)
counter=counter+1;
        end of loop
end of loop
```

//if node having no outlinks then equally distribute probability

If a row of probability Transition Matrix has no 1's, then replace each element by 1/N otherwise distribute it according to no of out links

```
if(counter==0) then
        for j=0 to N do
                prob[i][j]=1/N;
        end of loop
else
        for j=0 to N do
                if(prob[i][j]==1) then
                prob[i][j]=1.0/counter;
                end if
        end loop
end if else
```

Multiply the resulting matrix by 1 – Moister Factor.

Add Moister Factor/N to every entry of the resulting matrix, to obtain a Probability Transition Matrix.

```
For i=0 to N do
   For j=0 to N do
prob[i][j]=(prob[i][j]*(1-m _ fact))+((m _ fact)/N);
   end of loop
end of loop
```

Randomly select a node from 0 to N to start a walk

```
s _ int=random(N);
```

Initialize Random surfer and itr to keep account of number of iterations required.

```
x[s _ int]=1.0;
itr=0;
```

Try to reach at steady state within 200 iterations otherwise toggling occur

```
increment the itr
Itr=itr+1;
```

Multiplying probability transition matrix with probability vector to get steady state.

```
For i=0 to N do
```

```
            set  mult[i]=0;
            For  j=0  to  N  do
            mult[i]=mult[i]+x[j]*prob[j][i];
     end of loops
```
Check either system enter in steady state or not
```
For i=0 to N do
                if(mult[i]!=x[i]) then
                start  a  new  iteration
                end if
        else
        print  the  ranks  stored  in  Probability  vector  X  and
        goto 14.
          end loop
exit
end;
```

This ranking in this algorithm is done using the four modules of the algorithm that is proposed in this thesis; details are mentioned here:

- **a. getnodes():** This module is the member function of class WebDocumentRank, which is publicly declared. Its basic use is to construct a node, i.e. dummy document, in the collection and keep an account of how many documents are there in the collection. It basically scans the name of the document and assigns a unique document identity (docid) to every node starting from 1.
- **b. getlinks():** It is used to construct a link between two nodes. It is basically used to simulate the behavior of the hyperlink structure of the web. It first scans the name of the source node and then checks its validity, whether it exists or not, and if it exists, then demands the destination node again check its validity, and if both exist, then stores the information of that link into the adjacency matrix.
- **c. calrank():** This module is the member function of class WebDocumentRank, which is publicly declared. This module gets an input parameter from its caller [26,27] using a call by value concept named Moister Factor, denoted by m_fact. Then it will set up the probability transition matrix by adding the concept of random surfer model and Markov chain. After that, it will select a random node, from where it will start its walk using a random function. After that, this module tries to reach at a steady state within 200 iterations, and if the system is not able to reach in the steady state up to 200 iterations, then it is concluded that the system moves in the toggle state and can't be reached in the steady state. To reach the steady state, it multiplies the probability vector with the probability transition matrix. And the steady state is reached when the two successive probability vectors are the same. This module is also used to keep account of the number of iterations required to reach a steady state.
- **d. validity_check():** This module is the member function of class WebDocumentRank, which is publicly declared. This module gets an input parameter from its caller using a call by value concept named with the name of the node whose validity has to be checked denoted by node. This function

is called by getlinks() function to check whether the link information provided is valid or not, i.e. either the nodes exist between which user wants a link. If the node exists this module will return its ID; otherwise it returns -1.

3.1.1 Optimized Ranked Pages

The algorithm enables any search engine to rank the page depending on the number of inbound links to a particular page from an authentic source. The user that initially implemented the search gets the results that are relevant and have been linked and referred by other web pages.

4 RESULT ANALYSIS

This algorithm is implemented using the concept of Object Based [28,29] Language C++. In it, there is a class named WebDocumentRank, which includes various member functions to perform various operations like to create a node (dummy document) in the collection or create a link between documents that basically simulate the effect of hyperlink of one document to another [30,31], which actually uses a validity check function to ensure that before creating a link the nodes exist. One module is used to compute the WebDocumentRank on the basis of link structure.

4.1 Assumptions

- In the event there are different connections between two pages, just a solitary edge is set.
- No self-circles are permitted.
- The edges could be weighted; however, we expect that no weight is doled out to edges in the chart.
- Links inside a similar site are taken out on the grounds that they don't pass on an endorsement, utilized with the end goal of the route.
- Isolated hubs are eliminated from the diagram.

4.2 Input Web Graph 1

Figure 2.2 shows the input graph used for the calculations and result analysis.

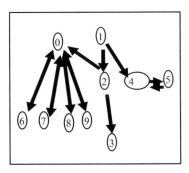

FIGURE 2.2 Web graph 1 used for analysis.

CASE 1: CONVERGENCE SPEED DECREASES AS THE MOISTER FACTOR INCREASES

A Data Collected for Analysis

To analyze the convergence speed, we collect the information on the number of iterations required by a random surfer to reach at a steady state. The input data is shown in Table 2.1.

B Graphical Observations

Figure 2.3 shows the graph between Moister Factor and number of iterations.

C Results

- Convergence speed decreases as the Moister Factor moves from 0 to 1.
- Therefore, the damping factor must be selected closer to 1 from the point of convergence speed.

TABLE 2.1
Moister Factor versus number of iterations

Moister Factor	No. of Iterations
0	Infinity
0.05	Infinity
0.1	159
0.15	105
0.2	Infinity
0.25	Infinity
0.3	61
0.35	Infinity
0.4	34
0.45	Infinity
0.5	33
0.55	24
0.6	23
0.65	22
0.7	18
0.75	18
0.8	18
0.85	12
0.9	12
0.95	10
1	2

Optimizing Web Page Ranks

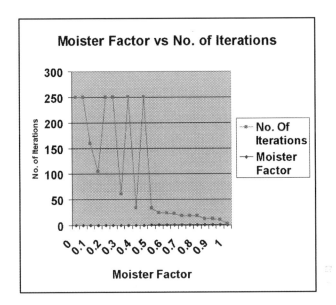

FIGURE 2.3 Graph between moister factor and number of iterations.

Note: 250 on the y-axis shows the infinity state.

TABLE 2.2
Ranks provided at Moister Factor 1

DOC ID	Rank Provided
1	.1
2	.1
3	.1
4	.1
5	.1
6	.1
7	.1
8	.1
9	.1
10	.1

CASE 2: AS THE MOISTER FACTOR IS 1, RANDOM SURFER ENTERS INTO THE IDEAL STATE

A Data Collected for Analysis

Table 2.2 contains the data for analysis at MF = 1. First column contains the document IDs and the second column contains the ranks provided at MF = 1.

B GRAPHICAL OBSERVATION

For analyzing the result against the data in Table 2.2, Figure 2.4 shows the graph between ranks provided to different documents with different Moister Factor.

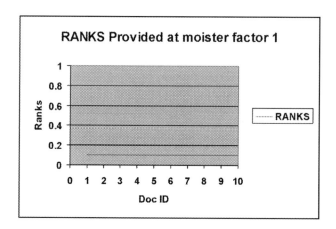

FIGURE 2.4 Graph between ranks provided to different documents at moister factor.

C RESULT

The Markov process enters into the ideal state when the Moister Factor is 1, as all the documents get the same rank independent of their structure.

CASE 3: AS THE MOISTER FACTOR IS LESS THAN 1, RANDOM SURFER ENTERS INTO THE IDEAL STATE

Random surfer can enter into the toggle state when the Moister Factor is less than 0.5.

A DATA COLLECTED FOR ANALYSIS

Table 2.3 contains the data for Moister Factor less than .5 and the ranks given to document when Moister Factor is less than .5.

TABLE 2.3
Ranks provided at different Moister Factors (< .5) to different documents

Doc ID	Moister Factor	Ranks Provided
1	0	∞
2	0	∞
3	0	∞
4	0	∞
5	0	∞
6	0	∞

Optimizing Web Page Ranks

Doc ID	Moister Factor	Ranks Provided
7	0	∞
8	0	∞
9	0	∞
10	0	∞
1	.05	∞
2	.05	∞
3	.05	∞
4	.05	∞
5	.05	∞
6	.05	∞
7	.05	∞
8	.05	∞
9	.05	∞
10	.05	∞
1	.1	0.20894
2	.1	0.050161
3	.1	0.035125
4	.1	0.028359
5	.1	0.244327
6	.1	0.232446
7	.1	0.050161
8	.1	0.050161
9	.1	0.050161
10	.1	0.050161
1	.15	0.231153
2	.15	0.057365
3	.15	0.04245
4	.15	0.036111
5	.15	0.20832
6	.15	0.195141
7	.15	0.057365
8	.15	0.057365
9	.15	0.057365
10	.15	0.057365
1	.2	∞
2	.2	∞
3	.2	∞
4	.2	∞
5	.2	∞
6	.2	∞
7	.2	∞
8	.2	∞
9	.2	∞
10	.2	∞
1	.25	0.244542
2	.25	0.065324
3	.25	0.053139
4	.25	0.04857
5	.25	0.170563

(*Continued*)

TABLE 2.3
Ranks provided at different Moister Factors (< .5) to different documents

Doc ID	Moister Factor	Ranks Provided
6	.25	0.156565
7	.25	0.065324
8	.25	0.065324
9	.25	0.065324
10	.25	0.065324
1	.3	∞
2	.3	∞
3	.3	∞
4	.3	∞
5	.3	∞
6	.3	∞
7	.3	∞
8	.3	∞
9	.3	∞
10	.3	∞
1	.35	∞
2	.35	∞
3	.35	∞
4	.35	∞
5	.35	∞
6	.35	∞
7	.35	∞
8	.35	∞
9	.35	∞
10	.35	∞
1	.4	.236777
2	.4	.072220
3	.4	.065473
4	.4	.063449
5	.4	.143371
6	.4	.129829
7	.4	.072220
8	.4	.072220
9	.4	.072220
10	.4	.072220
1	.45	∞
2	.45	∞
3	.45	∞
4	.45	∞
5	.45	∞
6	.45	∞
7	.45	∞
8	.45	∞
9	.45	∞
10	.45	∞

Optimizing Web Page Ranks 31

B GRAPHICAL OBSERVATION

For analyzing the result against the data in Table 2.3, Figure 2.5 shows the Ranks provided at different moister factor (< 0.5) to different documents.

C RESULTS

The filled area with pink in Figure 2.5 up to level 1 shows that no rank is provided, and the system goes into toggling state.

4.3 OVERALL RESULT ANALYSIS AND PERFORMANCE

These results are presented in the form of a performance graph showing that:

At **Moister Factor = 0.95**, the proposed algorithm ranks the web pages that were not able to secure a rank previously using comparative algorithms. Figure 2.6 shows the comparison and outperformance of the proposed work.

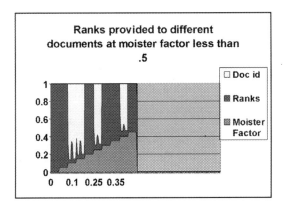

FIGURE 2.5 Ranks provided at different moister factor (< 0.5) to different documents.

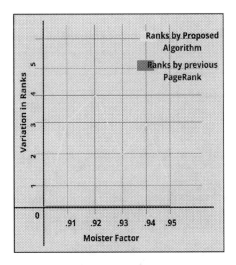

FIGURE 2.6 Performance: ranks provided by proposed algorithm vs. page rank algorithm.

Performance graph Figure 2.6 clearly shows that the previous page rank algorithm does not rank the pages on Moister Factor ranging from 0.91 to 0.95. However, with the proposed algorithm, the ranks of web pages can be obtained. Hence, the proposed algorithm outperforms the previous algorithms in terms of providing ranking on a range of Moister Factors.

5 CONCLUSION

This study depicts how the link structure [321/n37] of the web can be used to provide the ranking to various documents. The research carried out in the chapter proves that this ranking can be provided offline. With the help of our proposed algorithm, accurate and relevant prioritization of various documents on the web independent of the query is done. After the analysis it is concluded that Moister Factor must not be selected closer to zero. At Moister Factor 1, the system enters into the ideal state so that the ranking provided is insignificant. Hence, the Moister Factor must be selected greater than or equal to 0.5. However, if we consider convergence speed as the only factor to evaluate the performance, then the best Moister Factor **MF = 0.95**. The proposed algorithm is a query independent algorithm; it does not consider query during ranking, and it shows that a document is authorized if it links to other authentic documents that are again linked to several other relevant and recognized web pages.

REFERENCES

[1] Manning, C.D., P. Raghavan, H. Schütze. *Introduction to Information Retrieval*, vol. 1 (Cambridge University Press, Cambridge, 2008), p. 496.
[2] Brin, Sergey Lawrence Page. Reprint of: The anatomy of a large-scale hypertextual web search engine. *Computer Networks* 56(18), 3825–3833 (2012).
[3] Henzinger, Monika R., Rajeev Motwani, Craig Silverstein. Challenges in web search engines. *SIGIR Forum* 36(2), 11–22, 15 (2002).
[4] Chaudhri, A.B., A. Rashid, R. Zicari (eds.). *XML Data Management: Native XML and XML-Enabled Database Systems* (Addison Wesley, Boston, 2003).
[5] Maarek, Y., F. Smadja. Full text indexing based on lexical relations: An application: Software libraries. Proceedings of the 12th International ACM SIGIR Conference on Research and Development in Information Retrieval, New York City, NY, 1989: 198–206.
[6] Xing, Wenpu, Ali Ghorbani. "Weighted page rank algorithm": Communication networks and services research, 2004. Proceedings Second Annual Conference on. IEEE Fredericton, Canada, 2004.
[7] Tang, M., Y. Bian, F. Tao. The research of document retrieval systems based on the semantic vector space model. *J Intelligence* 5(29), 167–177 (2010).
[8] Yang, X., D. Yang, M. Yuan. Scientific literature retrieval model based on weighted term frequency. Intelligent Information Hiding and Multimedia Signal Processing (IIH-MSP), 2014 Tenth International Conference on. IEEE, 2014: 427–430.
[9] Yan, Y., J. Du, P. Yuan. Ontology-based intelligent information retrieval system. *Journal of Software* 26(7), 1675–1687 (2015).
[10] Ma, C., W. Liang, M. Zheng, H. Sharif. A connectivity-aware approximation algorithm for relay node placement in wireless sensor networks[J]. *IEEE Sensors Journal* 16(2), 515–528 (2016).

11. Xu, M., Q. Yang, K.S. Kwak. Distributed topology control with lifetime extension based on non-cooperative game for wireless sensor networks[J]. *IEEE Sensors Journal* 16(9), 3332–3342 (2016).
12. Vallet, D., M. Fernández, P. Castells. An ontology-based information retrieval model. *The Semantic Web: Research and Applications* (Springer, Berlin Heidelberg, 2005), pp. 455–470.
13. Baeza-Yates, R., G. Navarro. Integrating contents and structure in text retrieval. *SIGMOD Record* 25(1), 67–79 (1996).
14. Tirri, H. Search in vain: Challenges for internet search. IEEE Computer, Elsevier, Amsterdam, The Netherlands, 2003 Jan: 115–116.
15. Kleinberg, J.M. Authoritative sources in a hyperlinked environment. *Journal of the ACM (JACM)* 46(5), 604–632 (1999).
16. Kumar, Gyanendra, Neelam Duhan, A.K. Sharma. Page Ranking based on number of visits of links on a Web page. Computer and Communication Technology (ICCCT), 2011 2nd International Conference on. IEEE, Allahabad, India, 2011.
17. Tyagi, Neelam, Simple Sharma. Weighted Page Rank algorithm based on number of visits to links of web pages. *International Journal of Soft Computing and Engineering (IJSCE)* 2231–2307 (2012).
18. Zhang, Y., K. Nan, Y. Ma. Research on ontology-based information retrieval system models. *EURASIP Journal on Wireless Communications and Networking* 8(25), 2241–2249 (2008).
19. Lu, Y., M. Liang. Improvement of text feature extraction with genetic algorithm. *New Technology of Library and Information Service* 4(245), 48–57 (2014).
20. Premalatha, R., S. Srinivasan. Text processing in information retrieval system using vector space model. Information Communication and Embedded Systems (ICICES), 2014, International Conference on. IEEE, Chennai, India, 1–6.
21. Jones, K.S., S. Walker, S.E. Robertson. A probabilistic model of information retrieval: Development and comparative experiments: Part 1. *Information Processing and Management*. 36(6), 779–808 (2000).
22. Wong, S.M., W. Ziarko, P.C. Wong. Generalized vector spaces model in information retrieval. Proceedings of the 8th Annual International ACM SIGIR Conference on Research and Development in Information Retrieval. ACM, Montreal Quebec Canada, 1985: 18–25.
23. Castells, P., M. Fernández, D. Vallet, P. Mylonas, Y. Avrithis. *On the Move to Meaningful Internet Systems 2005: OTM 2005 Workshops: Self-Tuning Personalized Information Retrieval in an Ontology-Based Framework* (Springer, Berlin Heidelberg, 2005), pp. 977–986.
24. Pereira, R.M., A. Molinari, G. Pasi. Contextual weighted representations and indexing models for the retrieval of HTML documents. *Software Computation Journal* 9(7), 481–492 (2005).
25. Keßler, C., M. Raubal, C. Wosniok. *Smart Sensing and Context: Semantic Rules for Context-Aware Geographical Information Retrieval* (Springer, Berlin Heidelberg, 2009), pp. 77–92.
26. Cleverdon, C.W. On the inverse relationship of recall and precision. *Journal of Documentation* 23, 195–201 (1972).
27. World Wide Web Consortium (W3C). *On SGML and HTML*, (n.d.). Retrieved February 8, 2004 from.
28. Bulman, D.M. An object-based development model: Computer language. *HP Professional Books* (Electrical and Computer Engineering Publications, Ontario, Canada, 2007).

[29] Banerjee, J., H.-T. Chou, J.F. Gm, W. Kim, D. Woelk, N. Ballou. Data model issues of object-oriented applications. *ACM Transactions on Office Information Systems* 5(1) (1987).

[30] Arasu, A., J. Cho, H. Garcia-Molina, S. Raghavan. Searching the web. *ACM Transactions on Internet Technologies* 1 (1) (2001).

[31] Baeza-Yates, R., B. Ribeiro-Neto. *Modern Information Retrieval* (Addison-Wesley, Harlow, England, 1999).

[32] Baeza-Yates, R., F. Saint-Jean, C. Castillo. *Web Dynamics, Structure and Page Ranking, SPIRE 2002* (Springer LNCS, Lisbon, Portugal, 2002), pp. 117–130.

[33] Baeza-Yates, R., J. Piquer. *Agents, Crawlers, and Web Retrieval, CIA 2002* (Springer LNIA, Madrid, Spain, 2002), pp. 1–9.

[34] Baeza-Yates, R., D. Carmel, Y. Maarek, A. Sofer (eds.). IR and XML. *JASIST* 53(6) (2002) (Special issue).

[35] Baeza-Yates, R., B. Ribeiro-Neto. *Modern Information Retrieval*, vol. 463 (ACM Press, New York, 1999).

[36] Voorhees, E.M., D.K. Harman (eds.). *TREC: Experiment and Evaluation in Information Retrieval*, vol. 1 (MIT Press, Cambridge, 2005).

[37] Cao, Y.G., Y.Z. Cao, M.Z. Jin, C. Liu. Information retrieval oriented adaptive Chinese word segmentation system. *Journal of Software* 3, 17 (2006).

3 Design and Implementation of Novel Techniques for Content-Based Ranking of Web Documents

Ayushi Prakash, Sandeep Kumar Gupta, & Mukesh Rawat

CONTENTS

1 Introduction 35
   1.1 Search Engine 36
      1.1.1 Crawl on Web 36
      1.1.2 Database 37
      1.1.3 Search Interface 37
   1.2 Structure of SE (Search Engine) 37
      1.2.1 Predefined Keywords 37
      1.2.2 Generation of Inverted Index 37
2 Architecture Design for Content-Based Ranking of Web Documents 38
   2.1 Model Description 38
      2.1.1 Query Preprocessing 38
3 Ranking Parameters for Organizing the Information within Collection 43
   3.1 Precision 43
   3.2 Recall 44
   3.2 F-Measure 44
   3.3 Term Proximity 44
4 Conclusion 45
References 45

1 INTRODUCTION

Information retrieval: A search engine is basically software, or we can say a system of organization which is mapped out to find out the WWW or internet cast about in structured and in an organized way, with the help of a query engine. A search engine is made up of tools that are based on the web that enables different types of users to get information on the World Wide Web. Search engines like Yahoo, Google, MSN, etc. are helpful if users want to get relevant information on the web. Generally, a search engine uses different mathematical formulae to get the appropriate results. Apart from mathematical formula, search engines use different algorithms and, with the help of these algorithms that work basically on key elements of web pages, as for example titles, content and words density shows ranking for where the results will be placed on the page. For making results more relevant, these algorithms are modified and revised constantly.

Information retrieval: Accessing the right and relevant information in less time, or we can say fast searching of relevant data, is very important nowadays and is the way by which we can get the desired information or access the relevant information, known as information retrieval. There are some recovery programs, also known as information recovery programs that part of accessing information on the web. Getting the right information related to the relevant topic from a large database containing all kind of data, metadata, directories, indexes, etc., is known as retrieving information. Search depends on different criteria, sometimes based on metadata or sometimes on the whole text as well. Elements of search engine are shown in Figure 3.1. We show this process as follows.

1.1 SEARCH ENGINE

A web search engine is created to find information on the WWW. The search outcomes are normally given out in a line of results, frequently referred to as search engine results pages (SERPs). The information may be a proficient in web pages, images, information and other types of files. Some Search Engines (SE) also excavates data available in databases or open directories. Unlike web directories, which are maintained only by human editors, SE also maintains real-time information by running an Algorithm on a web crawled.

A search engine operates in the following manner:

- Crawl on web
- Indexing of web
- Search interface (SI)

1.1.1 Crawl on Web

The software or a program main responsible to browse the web by browsing the documents automatically. Web Crawler alias Spider, index the content of websites to provide up to date data. Crawling on web refers crawling from one page to another relevant pages through hyperlinks to validate the required and relevant data.

Database ⟹ USER ⟹ User's Query ⟹ Search Mechanism Dissemination

FIGURE 3.1 Elements of a search engine: We can define the elements in three parts, broadly.

1.1.2 Database

To manage, access, and update the collection of information in an organized way in a computer system, we use a systematic data structure known as a database, a reservoir or virtual depository consisting of huge web resources such as pictures, videos, audios, and numerous different files and records.

1.1.3 Search Interface

Generally, an interface is a device or system used to interact between two unrelated entities or objects. A search engine is also the interface that acts as a link between the end user and the database reservoir.

1.2 STRUCTURE OF SE (SEARCH ENGINE)

In lieu of searching all documents, the search engine explores the existing database to find the catchword [4]. Afterwards, by use of software, it searches the requested information stored in the database with the help of a web crawler. That is composed of two processors a front-end processor (FEP) and a back-end processor (BEP).

There are two main kinds of SE that have been developed.

1.2.1 Predefined Keywords

A system where we keep the predefined and relevant words that human beings have programmed widely to optimize the search engine is known as Predefined Keywords system.

1.2.2 Generation of Inverted Index

Another system that produce an "Inverted Index" by exploring and observing texts it placed is known as generation of Inverted Index System. This system depends more specifically on the computer itself to do the maximum work. Generation of inverted index is shown in Figure 3.2.

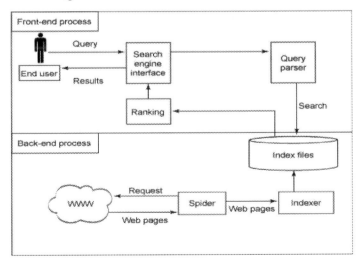

FIGURE 3.2 Generation of inverted index.

We can take overview of the front-end and back-end process with the help of Figure 3.2.

2 ARCHITECTURE DESIGN FOR CONTENT-BASED RANKING OF WEB DOCUMENTS

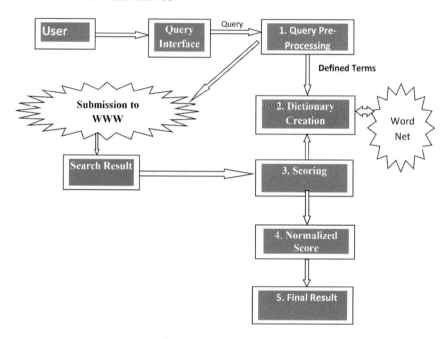

FIGURE 3.3 Work flow of content-based ranking of web documents.

2.1 Model Description

2.1.1 Query Preprocessing

In the proposed model, we have given a novel way to identify and search the optimized form of the term to be indexed to achieve the best retrieval performance, with the help of query preprocessing on the query given by USER through the query interface.

Following are the stepwise tasks involved in query preprocessing:

i. Tokenization/segmentation using Python
ii. Stemming and lemmatizer
iii. Parts of speech
iv. Parsing
v. Context analysis
vi. Sentiment analysis

Design and Implementation of Novel Techniques

2.1.1.1 Tokenization/ Segmentation

Tokenization or segmentation is a process of generating tokens from a query string. A process is required to split the query string into all list of segments; these segments known as tokens. Tokenization is splitting of the string into tokens with a basic objective to make string easy and convenient. String is the uncomplicated and easiest method to represent the data, but it is very difficult to interpret this format. So it is important to bifurcate the string into smaller chunks.

This can be done in a number of ways: The splitting of a given string containing text into different sentences or splitting based on individual words.

Sentence Tokenization Sentence tokenization is a process when the splitting breaks the string into sentences. For example, if the string is "Hello NLP Learner. This is an NLP Python Script,"after doing sentence tokenization, the output will be:

"Hello NLP Learner,""This is an NLP Python Script."

Here, we can see how the single string splits into two different sentences to make string easy and convenient. Sentence tokenization is shown in Figure 3.4.

Word Tokenization Word tokenization is a process of splitting the given string into words to make strings easy and convenient. For example, "We are learning artificial intelligence." Here, by applying word tokenization to the string, we get: "We" "are" "learning" "artificial" "intelligence." Finally, one string splits into five words. Word tokenization is shown in Figure 3.5.

Algorithm for Tokenization

```
fromnltk.tokenize import sent_tokenize
sentence=sent_tokenize("Hellow    world.    world    has
   become  different  now !  It  is  becomming difficult day
   by day")
print(sentence)
```

- Sentence tokenizing using NLTK and python

    >>> from nltk.tokenize import sent_tokenize

    >>> sentence = sent_tokenize("Hello NLP learner. This is an NLP python script")

    >>> print (sentence)

FIGURE 3.4   Illustration of sentence tokenization.

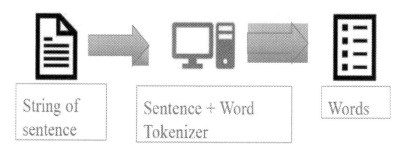

- Word tokenization using NLTK & Python

>>> from nltk.tokenize import word_tokenize
>>> words = word_tokenize("We are learning Artificial Intelligence")
>>> print (words)
>>> ['We', 'are', 'learning', 'Artificial', 'Intelligence']

FIGURE 3.5 Illustration of word tokenization.

TABLE 3.1
Difference between stemming and lemmatization

Word	Stemming	Lemmatization
information	Inform	information
informative	Inform	informative
computers	Comput	computer
feet	feet	foot

i Stemming and Lemmatizer

To minimize the inflection form, stemming is the process to generate root words from inflected words; the generated words are a derived form to deduce it to the common base form. The lemmatizer generates the root words from the actual word, but in stemming, the stem might not be an actual word. In a normalization process, each has its own equal importance at the time of canonical representation for a set of related words. The difference between stemming and lemmatization is shown in Table 3.1.

Stemming It is a process to lop off the word endings in order to achieve the objective. It includes removal of derivational affixes. This is the process used by many query search engines to analyze the meaning behind the word. It is fast because this method does not stick with the context of the word. Example of Stemmed words is shown in Table 3.2.

For example, shortlisted, shortlisting, shortlister ⟹ shortlist

Design and Implementation of Novel Techniques

TABLE 3.2
Example of stemmed words

	Word	Stemmed Word
0	Shortlist	Shortlist
1	shortlisting	Shortlist
2	Shortlisted	Shortlist
3	Shortlister	Shortlist

TABLE 3.3
Illustration of words and their stemming and lemmatization

Word	Stemming	Lemmatization
information	inform	information
informative	inform	informative
computers	comput	computer
feet	feet	foot

Here we can see all the words related to the base words can be stemmed.

Algorithm for Stemming

```
fromnltk.stem import SnowballStemmer
fromnltk.tokenize import word_tokenize
ks=SnowballStemmer("english")
list_words1=["python","pythoner","pythoning","pythoned",
   "pythonly","wolves", "leaves","horses","dogs","fairly"]
fori in list_words1: print(ks.stem(i))
```

Lemmatizer Lemmatizing is process in which the main concern is the name reduction to improve the quality in comparison with the stemmer. It contains function that uses an all-inclusive full-form dictionary, which enables it to deal with irregular forms. It also provides a substantially higher-quality and high-cost version of the stemmer. Illustration of words and their stemming and lemmatization are shown in Table 3.3.

For example: Women ⟹ woman, wolves ⟹ wolf

Algorithm for Lemmatizing

```
From nltk import stem
lsm=stem.WordNetLemmatizer()
list_words2=["running",    "Jumping"    ,
"Skipping","knives","wives","Pythonly",
"Pythony","Geese","dogs","women"]
fori in list_words2:
print(lsm.lemmatize(i))
```

Dictionary Creation Dictionary creation is a process to store the data from the WordNet, to identify the relevant words related to the query strings. WordNet is a virtual depository for the English language, which was created by Princeton and is part of the NLTK corpus. WordNet, alongside the NLTK module, is used to find the definition and explanation of words, synonyms, antonyms, and more. Let's cover some examples. Example of dictionary creation is shown in Table 3.4.

Much natural-language processing requires a large lexical database of word relations. The dictionary has much more consistent structures that can be exploited by a computer to recognize word relations. Dictionary creation may also contribute to the identification of definite groups of relation instances that are difficult to search from general texts.

Bring in WordNet:
Use of word "process" to get synsets here:

```
syns=wordnet.synsets("process")
with a sample _ synset:
print(syns[0].name())
plan.n.01
see here
print(syns[0].lemmas()[0].name())
plan
Explanation _ synset:
print(syns[0].definition())
To get the results, we use series of instructions
Specimen of the term
print(syns[0].examples())
Now, how we detect synonyms and antonyms to a word?
  Find the list of synonyms and antonyms:
Synonyms _ 11=[]
Antonyms _ 1=[]
Forsyninwordnet.synsets("beautiful"):
forLinsyn.lemmas():
synonyms.append(l.name())
if L antonyms():
antonyms.append(l.antonyms()[0].name())
print(set(synonyms _ 1))
print(set(antonyms _ 1))
```

TABLE 3.4
Example of dictionary creation

Synonyms	beneficial	upright	charmingly	elegantly
Antonyms	shoddy	average	bad	weak

Design and Implementation of Novel Techniques 43

Algorithm for Calculating the Score and Normalized [3] *Score of Search Results* In the following algorithm, keywords of the search result documents are matched with each of the dictionary words. If the keyword is found in the dictionary list, its score is calculated in the form of total occurrences of that particular keyword in the search document. But this procedure gives weightage to the word which is most frequently occurring in the document.

To normalize these things, a normalized score is calculated in which each calculated score of the term of the document is divided by the average score of the whole search document. And new ranking of the search documents is calculated by arranging the search documents according to their decreasing ranks.

Input: Search results

S. No.	Doc ID	URL
1.	SR_1	https://en.wikipedia.org/wiki/Op-system
2.	SR_2	https://whatis.techtarget.com/definition/ML-System-AI
3.	SR_3	https://edu.gcfglobal.org/en/computerbasics/understanding-ML
4.	SR_4	www.tutorialspoint.com/operating_system/os_Intro.htm

```
D = {k₁, k₂,..., kₙ}, where D is a dictionary
Step 1: S = { Kᵢ where Kᵢ ∈ SRⱼ , 1<i<n, 1<j<N}
n is the n.o. of keywords in jth document.
N is the n.o. of search results.
Step 2: for j=1 to N
            For i=1 to n
                If (kᵢ present in D)
                    Score (Kᵢ) = no. of occurrences
                        of Ki in SRⱼ
SCⱼ+=score(kᵢ)
AV(SCⱼ) = SCⱼ/n
Step 3: for j=1 to N
            For i=1 to n
                If(kᵢ present in D)
                    N-score(ki) = score(kᵢ)/ AV(SCⱼ)
New(SCⱼ)+=N-score(kᵢ)

Step 4: Arrange (SR₁, SR₂,..., SRⱼ,... SRₙ) in descending order
        of their calculated
        Normalized new score.
```

3 RANKING PARAMETERS FOR ORGANIZING THE INFORMATION WITHIN COLLECTION

3.1 PRECISION

Precision (P) is the fraction of the files retrieved which can be applicable to the user's facts need. It takes all retrieved files into account [2]. It also can be evaluated at a given cutoff rank that specifies the relevant pages identified by the proposed system.

$$\text{Precision} = \frac{\{\text{no. of applicable files}\} \cap \{\text{no. of retrieved files}\}}{\{\text{no. of retrieved files}\}}$$

3.2 RECALL

Recall (R) is the fraction of the files which are applicable to the question which are correctly retrieved. It may be checked out because this shows the irrelevant files retrieved by the proposed system [2].

$$\text{Recall} = \frac{\{\text{non-applicable files}\} \cap \{\text{retrieved files}\}}{\{\text{retrieved files}\}}$$

3.2 F-MEASURE

F-measure is basically to combine the value of precision and the value of Recall into a single measure that captures both properties.

The traditional F-measure is calculated as follows:

$$F - \text{measure} = \frac{(2*P*R)}{(P+R)}$$

3.3 TERM PROXIMITY

Term proximity is a structure of the dependency of terms based on the distance in a given document. An information retrieval system using term proximity allocates a higher score to documents in which the query terms (QT) are near each other. The method to find the measure of similarity between terms given in the document selected. Let us assume an example of string and terms [1].

The query string: A, D, T
Selected document string: P,Q,A,X,D,Y,T,U,Z,A,B,D,T,Z

1	2	3	4	5	6	7	8	9	10	11	12	13	14
P	Q	A	X	D	Y	T	U	Z	A	B	D	T	Z
Distance		⇑		⇑		⇑			⇑		⇑	⇑	
		3		5		7			10		12	13	

We can see the distance between the terms in the given document. It can be calculated with the help of term proximity, so by this way, the term proximity declares the high scores to those documents where the QT are near each other.

4 CONCLUSION

In this chapter, we proposed a novel technique and an algorithm of normalization by which we get average score despite the high frequency of irrelevant or less important words. To normalize these things, a normalized score is calculated in which each calculated score of term of the document is divided by the average score of the whole search document. And new ranking of the search documents is calculated by arranging the search documents according to their decreasing ranks. By this method, the average performance of ranking of keywords improves its effectiveness and competence for different search engines. This method has the benefits of being appropriate in a variety of contexts and trying to provide more relevant and important words in comparison to earlier algorithms. Our method is not complicated for use and does not have any computational problem as compared to other methods. It is significant and effective and provides significant and powerful search in the context of the web.

In the fields of data mining and machine learning, rank aggregation with similarity has as its robust scope a method of unsupervised ML techniques such as KNN, CNN, and ANN. We can measure the performance of one proposed model, which tells that accuracy in the form of similarity between the search results. Future work will lie in further search methods in which this objective can be simultaneously achieved.

REFERENCES

[1] Amitay, E., Carmel, D., Lempel, R., & Soer, A., Scaling IR-system evaluation using term relevance sets. Proceedings of the 27th ACMSIGIR Conference, New York, NY, 2004, pp. 10–17.
[2] Hussain, Mohd, Prakash, Ayushi, Khan, Afsaruddin, A novel approach to automatically combine search and ranking results. Proceeding of IJCA on National Conference on Advancement of Technologies-Information Retrieval & Computer Network (ISCON), Foundation of Computer Science, New York, NY, 2012.
[3] Bruce, Croft W., Metzler, D., & Strohman, T., *Search engines informational retrieval in practice*. Pearson Education, London, 2015, pp. 25–26.
[4] Jain, Nandnee, Dwivedi, Upendra, Ranking web pages based on user interaction time. International Conference on Advances in Computer Engineering and Applications, IEEE Xplore, Ghaziabad, India, March 19–20, 2015, pp. 35–41.
[5] Cleverdon, C.W., The Cranfieldtests on indexing language devices. *Aslib Proceedings*, 19, 1967, 173–192.

4 Web-Based Credit Card Allocation System Using Machine Learning

Vipul Shahi, Yashasvi Srivastava, Utkarsh Sangam, Akarshit Rai, & Mala Saraswat

CONTENTS

1 Introduction 47
2 Literature Survey 48
3 Different Approaches for Allocation System 48
    3.1 Logistic Regression (LR Classification Algorithm) 48
    3.2 Random Forest Classifier 49
    3.3 K Neighbors Classifier 50
4 Proposed Methodology of System 50
5 Experiments and Results 52
    5.1 Description of the Data Set 53
    5.2 Null Values in the Data Set 54
    5.3 Target Variables in the Data Set 54
6 Results and Conclusions 55
References 56

1 INTRODUCTION

As we can see in our banks whenever someone is trying to get a credit card, they have to follow a long procedure. For example, if someone wants to get a credit card, they need to check whether they are eligible for the credit card or not and then if costumer is eligible for the credit card, they should apply for it and follow a long procedure for the allocation of the credit card.

The main objective of our project is to reduce the time spent by an individual who is applying for a credit card. So we are reducing this time and effort using a machine learning classification algorithm such as logistic regression. Now no one wants to wait for hours for eligibility for a credit card. No one wants to wait for the credit card to reach their home after one month. All this work can be done using a machine learning classification algorithm. This classification algorithm will tell whether an individual is eligible for the credit card, and if they are eligible, it will allocate the credit card to that individual. We will create a full-fledged application in which user information will be taken. Based on user entries, prediction will be made whether to allocate or not allocate credit card. If the person is eligible, a virtual credit card is allocated, and if the person is not eligible, the message is shown that "you are not eligible."

2　LITERATURE SURVEY

Khare et al. presented the moderating influence of multi-item list of value on credit card attributes in credit use among Indian customers. The research examines the impact of "lifestyle" such as convenience, use patterns, and status variables on credit card use. The study does not examine the influence of customer income, occupation, and education on credit card use, as many customers were not willing to disclose the information [1].

The research paper "Credit Cards and about Pre-Paid Cards" describes the characteristics of closed-system and open-system prepaid cards. Of particular interest is a class of open-system programs that offer a set of features similar to conventional deposit accounts using card-based payment applications. Unlike a debit card, a prepaid card is not linked to a bank account. Generally, when you use a prepaid card, you are spending money that you have already loaded onto the card [2]. Linda Mary Simon presented that the value of the card issuer is often related to the customer usage of the card or to the customer's financial worth. A majority of the respondents hesitated to comment on the statement "Plastic money leads to debt trap." The suggestions in this chapter are implemented to improve the services to the customer. This study seeks to know the extent of credit facilities available to the customers in various activities and their satisfaction towards such facilities [3]. The suggestions in this chapter are implemented properly to improve the services to the customer. John Leston's primary research objective of the consumer survey was to enhance Financial Conduct Authority's understanding of consumer behaviour in relation to shopping around and switching for credit cards [4].

3　DIFFERENT APPROACHES FOR ALLOCATION SYSTEM

3.1　Logistic Regression (LR Classification Algorithm)

Logistic regression is a supervised learning classification algorithm which is used to predict the probability of a target variable. The nature of the target or dependent variable is dichotomous, which means there would be only two possible classes (either 0 or 1) as shown in Figure 4.1.

This uses a sigmoidal function, which is:

$$S(z) = 1/(1+e-z) \qquad (4.1)$$

where $S(z)$ = output in 0–1(probability estimate)
z = input (your algorithm's prediction, e.g., $mx + b$)
e = base of natural log

Web-Based Credit Card Allocation

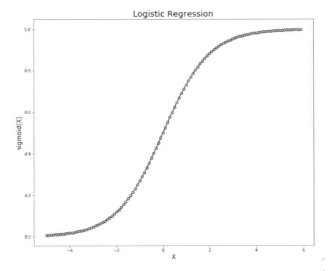

FIGURE 4.1 Logistic regression.

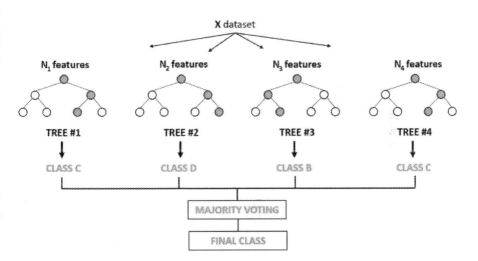

FIGURE 4.2 Random forest classifier.

3.2 RANDOM FOREST CLASSIFIER

Random forests or random decision forests are an ensemble learning method for classification, regression, and other tasks that operates by constructing a multitude of decision trees at training time. As the name suggests, it consists of a large number of individual decision trees that operate as an ensemble. Each individual tree in the random forest spits out a class prediction, and the class with the most votes becomes our model's prediction as shown in Figure 4.2.

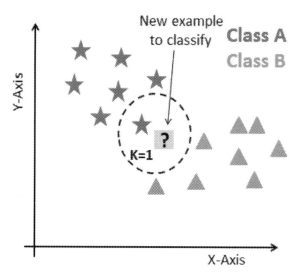

FIGURE 4.3 K Neighbors classifier.

3.3 K Neighbors Classifier

K Nearest Neighbor (KNN) is a very simple, easy-to-understand, versatile, and topmost machine learning algorithm. KNN is used in a variety of applications such as finance, healthcare, political science, handwriting detection, image recognition, and video recognition. In credit ratings, financial institutes will predict the credit rating of customers.

In KNN, K is the number of nearest neighbors. The number of neighbors is the core deciding factor. Suppose P1 is the point for which label needs to predict. First, you find the one closest point to P1 and then the label of the nearest point assigned to P1.

4 PROPOSED METHODOLOGY OF SYSTEM

The methodology that we have used for this system is a combination of machine learning (random forest classification algorithm) for predicting the eligibility and for the allocation of virtual credit card, and we have used various web development frameworks like HTML, CSS, and JavaScript. The flowchart for the proposed methodology is shown in Figure 4.4.

1. First, we will take the data of an individual who applied for the credit card
2. We apply classification algorithm to that data, and after applying the algorithm, the data is verified
3. If the individual is eligible, then we will allocate a virtual credit card to that individual
4. If the individual is not satisfying the criteria, we will notify him by displaying "not eligible." In the same way, web development frameworks will be used like HTML, CSS, and JavaScript

Web-Based Credit Card Allocation 51

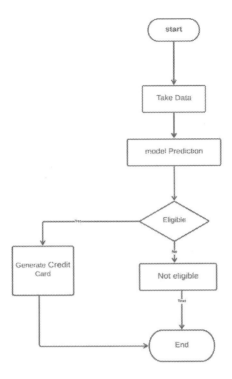

FIGURE 4.4 Flowchart.

The First Screen/Home page, where the user will enter the details to check their eligibility, is designed as shown in Figure 4.5.

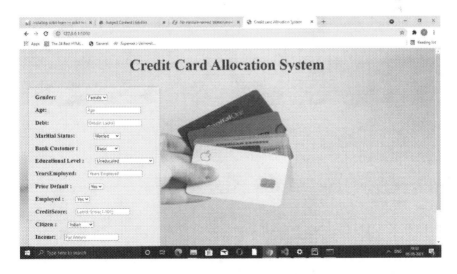

FIGURE 4.5 Home page.

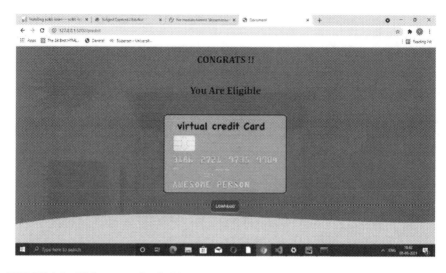

FIGURE 4.6 If the person is eligible.

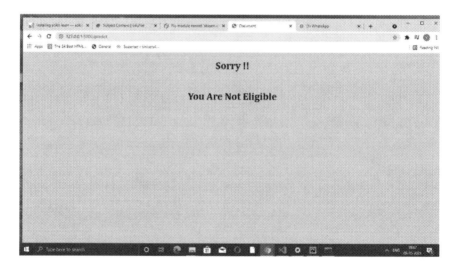

FIGURE 4.7 If the person is not eligible.

Figure 4.6 shows the screenshot if the person is eligible as per given data. Figure 4.7 depicts a screenshot if the person is not eligible.

5 EXPERIMENTS AND RESULTS

The web application of this system has a user-focused graphical user interface (GUI) to provide ease of use and effectiveness to users with different roles. The first page of the GUI system takes the relatable data input from the user to predict their eligibility for credit card based on the following user attributes:

1. Gender
2. Age
3. Debt
4. Marital Status
5. Bank Customer
6. Education Level
7. Ethnicity
8. Years Employed
9. Credit Score
10. Employed

5.1 Description of the Data Set

Figure 4.8 depicts the sample data set from Kaggle data set [6], and Figure 4.9 shows the analysis of the data.

```
In [2]: # Read the data file
        Data = pd.read_csv("CC_data.csv")
        Data
```

Out[2]:

	Male	Age	Debt	Married	BankCustomer	EducationLevel	Ethnicity	YearsEmployed	PriorDefault	Employed	CreditScore	DriversLicense	Citizen	Zi
0	b	30.83	0.000	u	g	w	v	1.25	t	t	1	f	g	2
1	a	58.67	4.460	u	g	q	h	3.04	t	t	6	f	g	4
2	a	24.5	0.500	u	g	q	h	1.50	t	f	0	f	g	2
3	b	27.83	1.540	u	g	w	v	3.75	t	t	5	t	g	1
4	b	20.17	5.625	u	g	w	v	1.71	t	f	0	f	s	1
...	...	...	...	...	...	...	...	...	...	...	...	...	...	...
685	b	21.08	10.085	y	p	e	h	1.25	f	f	0	f	g	2
686	a	22.67	0.750	u	g	c	v	2.00	f	t	2	t	g	2
687	a	25.25	13.500	y	p	ff	ff	2.00	f	t	1	t	g	2
688	b	17.92	0.205	u	g	aa	v	0.04	f	f	0	f	g	2
689	b	35	3.375	u	g	c	h	8.29	f	f	0	t	g	0

690 rows × 16 columns

FIGURE 4.8 Sample data set.

	Age	Debt	YearsEmployed	CreditScore	Income
count	678.000000	690.000000	690.000000	690.00000	690.000000
mean	31.568171	4.758725	2.223406	2.40000	1017.385507
std	11.957862	4.978163	3.346513	4.86294	5210.102598
min	13.750000	0.000000	0.000000	0.00000	0.000000
25%	22.602500	1.000000	0.165000	0.00000	0.000000
50%	28.460000	2.750000	1.000000	0.00000	5.000000
75%	38.230000	7.207500	2.625000	3.00000	395.500000
max	80.250000	28.000000	28.500000	67.00000	100000.000000

FIGURE 4.9 Description of data set.

5.2 Null Values in the Data Set

Figure 4.10 shows null values in the data set which are represented by white lines.

5.3 Target Variable in the Data Set

Figure 4.11 shows the target variable in which one variable is approved and other is non-approved.

`<AxesSubplot:>`

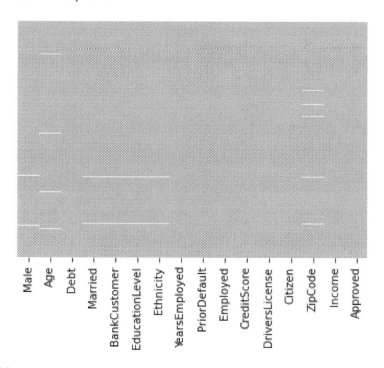

FIGURE 4.10 Null values in data set shown by the white line.

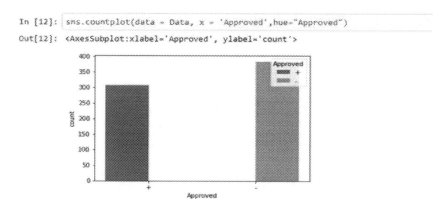

FIGURE 4.11 Target feature/variable.

Finally, after comparing the accuracy of various algorithms, we have come to a conclusion to use KNN in our project.

```
from sklearn.neighbors import KNeighborsClassifier

#Create KNN Classifier
knn = KNeighborsClassifier(n_neighbors=11)

#Train the model using the training sets
knn.fit(rescaledX_train, y_train)

#Predict the response for test dataset
y_pred = knn.predict(rescaledX_test)
```

```
from sklearn import metrics
print("Accuracy:",metrics.accuracy_score(Y_test, y_pred))
print("Precision:",metrics.precision_score(Y_test, y_pred))
print("Recall:",metrics.recall_score(Y_test, y_pred))
print("f_score:",metrics.f1_score(Y_test, y_pred))
```

Accuracy: 0.8840579710144928
Precision: 0.875
Recall: 0.9210526315789473
f_score: 0.8974358974358975

6 RESULTS AND CONCLUSIONS

Based on the results, we have reached two conclusions. The first one is when the person is eligible and the when the person is not. If the user is eligible, they get a virtual credit card, and if they are not eligible, they get a message saying, "Sorry, you are not eligible." From experiments, it is concluded that KNN provides a better result, with an accuracy of 88.4%. Figure 4.12 shows the comparison of various models such as random forest classifiers, logistic regression, Naive Bayes classifier, KNN, and decision tree classifier.

MODEL	ACCURACY	PRECISION	RECALL	F-MEASURE
Random Forest Classifier	86.2	82.7	94.7	88.3
Logistic Regression	85.5	86.8	86.8	86.8
Naive_bayes	83.3	76.7	100	86.8
KNeighbors Classifier	88.4	87.5	92.1	89.7
Decision Tree Classifier	78.9	77.6	86.8	81.9

FIGURE 4.12 Comparison of various models.

REFERENCES

[1] Khare, Arpita, Anshuman Khare, Shveta Singh, "'Factors affecting credit card use in India, March 2012': The research examines the impact of 'lifestyle' variables on credit card use", *Asia Pacific Journal of Marketing and Logistics.* 24(2):236–256. DOI: 10.1108/13555851211218048

[2] "Credit Cards and about Pre-paid Cards", Financial Consumer Agency of Canada. Archived from the original on 7 March 2007. Retrieved 9 January 2008. document: "Pre-paid Cards" (PDF). Financial Consumer. This paper describes the characteristics of closed-system and open-system prepaid cards.

[3] Simon, Linda Mary, "Customers satisfaction towards credit card services", *Indian Journal of Applied Research*, November 2012. The suggestions in this paper are implemented properly it to improve the services to the customer.

[4] Leston, John, "Credit Card Market Study: Consumer Survey", Prepared for the Financial Conduct Authority, September 2015. https://www.fca.org.uk/publication/market-studies/ms14-6-2-ccms-annex-3-1.pdf

[5] Kuhn, R., "Analysis of credit approval data", *Rstudio-pubs-static.s3.amazonaws.com*, 2020. Available at: http://rstudio-pubs-static.s3.amazonaws.com/73039_9946de135c0a49daa7a0a9eda4a67a72.html.

[6] Kaggle.com, *Default of Credit Card Clients Dataset.* Available at: www.kaggle.com/uciml/default-of-credit-card-clients-dataset.

[7] Khare, A. and Khare, A., "Factors affecting credit card use in India", [online] *Research Gate*, 2012. Available at: www.researchgate.net/publication/235319345 [Accessed 22 November 2020].

[8] Swami, S., "Payments in India going digital", [online] *Iaeme.com*, 2018. Available at: www.iaeme.com/MasterAdmin/UploadFolder/JOM_05_03_018/JOM_05_03_018:-pdf [Accessed 22 November 2020].

5 Pattern Recognition

Akanksha Toshniwal

CONTENTS

1 Introduction 57
2 Probability Theory..58
   2.1 Bayesian Probability Theory 59
3 Curve Fitting 60
   3.1 Bias and Variance 62
      3.1.1 Bias Variance Trade-Off 63
4 The Curse of Dimensionality 65
5 Classification..66
6 Clustering...68
Note...68
References..68

1 INTRODUCTION

Searching for patterns has a long history. For example, in the 16th century, Johannes Kepler discovered the empirical law of planetary motion based on the extensive astronomical observations by Tycho. The discovery of quantum physics established regularities in atomic spectra in the early 20th century. The human brain is programmed to observe regular patterns and reacts in the event of any irregularities. For example: a human baby can recognize a real elephant animal by mapping the features from the elephant drawings or cartoons movies. In computer science, the field of pattern recognition is concerned with the automatic discovery of regularities in data using computer algorithms and, with the use of these regularities, taking actions such as classifying the data into different categories (Bishop, 2006).

Consider an example such as recognizing categories of portraits, illustrated in Figure 5.1. Each image has 40 binary attributes and five landmark locations. Each image is represented by a vector of size 45, and the goal is to develop a machine algorithm which takes these feature vectors as input and produces the identification of classes like eyeglasses, wearing hat, wavy hair, bangs, moustache, smiling face, pointy nose, or oval face or to build a machine which performs face attribute recognition, face detection, landmark (or facial part) localization, and face editing and synthesis. Since humans have a wide variety of facial features, this is a non-trivial problem.

Pattern recognition has applications in fast-emerging areas like biometrics, bioinformatics, multimedia data analysis, machine learning, and data science, and it

DOI: 10.1201/9781003169550-5

FIGURE 5.1 Sample images from the CeleFaces data set. (From Liu, 2015.)

supports fields like computer vision, image processing, text and document analysis, and neural networks. There are two pattern recognition techniques. The first is supervised, which is used for classification and regression. The second is unsupervised, which is used for clustering. Besides these two techniques, the third one is called reinforcement learning or learning with critic, which learns binary signals (right or wrong) by feedback mechanism, e.g., in video gaming the scores are used as feedback.

2 PROBABILITY THEORY

Given a random experiment (i.e. pulling a card from a deck or rolling dice), a probability measures the population quantity that summarizes the randomness. Probability operates on the potential outcome from the experiment. For example: outcome of rolling a die is {1} or {1,2} or {2,4,6} or {1,3,5}, and so on. So probability is a function that takes any of these sets of possible outcomes and assigns a value between 0 and 1. To get an outcome between 1 and 6 on a die with probability = 1, the die must be rolled. The probability of any two sets of outcomes that are mutually exclusive is the sum of the respective probabilities. Example:

- *Let's say to compute the probability of the outcome 6 while rolling a single die once. The event $A = x_i$ (where $i = 1,2,3..6$), $p(A)$ = Probability of outcome A. $p(A = x_6) = 1/6 = 0.16$ or 16%. The probability of outcome not being 6 will be $p(A = x_{!6}) = 1 - p(A = x_6) = 0.84 = 84\%$.*

- Let's say the probability of the outcomes of a die roll is either {1,2} or {3,4} (mutually exclusive). Then for events A = $x_{\{1,2\}}$ and B = $x_{\{3,4\}}$, $p(A = x_{\{1,2\}}) = 0.33$ and $p(B = x_{\{3,4\}}) = 0.33$ so $p(A \cup B) = p(A) + p(B) = 0.66$ or 66%.
- Let's say the probability of outcome B = {3,4} after the outcome A = {1,2} on the event of die roll twice, then instances for which A = $x_{\{1,2\}}$ and the fraction of such instances for which B = $x_{\{3,4\}}$ is written $p(B = x_{\{3,4\}}| A = x_{\{1,2\}})$ and is called the conditional probability of B = $x_{\{3,4\}}$ given A = $x_{\{1,2\}}$, which gives product rule of probability i.e. $p(A,B) = p(B|A)p(A)$.

The Russian mathematician Kolmogorov proposed the following axioms for probability:

- The probability that nothing occurs is 0.
- The probability that something occurs is 1.
- If A and B are mutually exclusive outcomes, $p(A \cup B) = p(A) + p(B)$.
- For any event A, the probability of A is greater or equal to 0.

2.1 Bayesian Probability Theory

Bayesian probability theory was discovered by Thomas Bayes, who was an amateur scientist and a mathematician, in 1764 in his paper "Essay towards Solving a Problem in the Doctrine of Chances" (Dale, 2005). This approach is based on quantifying the trade-offs between various classification decisions using probability and the costs that accompany such decisions.

Imagine there are blue and red balls in a bag, and a dog is asked to fetch a ball from the bag. We would say that the next ball is equally likely to be blue or red. We assume that there is some a priori probability (prior) $P(A)$ that the next ball is blue and some a priori probability $P(B)$ that next ball is red. If there are only two colors, we can say that sum of probabilities is 1. The prior probability reflects prior knowledge about the total number of blue and red balls in the bag. If only this information is used to make the decision for the next fetch event, i.e. if $P(A) > P(B)$, decide A, otherwise B. But this solution works if it's a decision rule for the next ball fetch but the decision rule for all of the next fetches is odd. Let's say we show the dog what's in the bag for some time 'x' every time before fetch; we consider x to be a continuous random variable whose distribution depends on the state of environment and is expressed as $P(x|A)$. This is the class-conditional probability density function. We note first that the (joint) probability density of finding a pattern that is in category A or B and has feature value x can be written two ways: $p(A \text{ or } B, x) = P(A \text{ or } B \mid x) p(x) = p(x \mid A \text{ or } B) P(A \text{ or } B)$. Rearranging these leads us to the answer to Bayes's formula:

$$P(A \text{ or } B \mid x) = \frac{p(x \mid A \text{ or } B) P(A \text{ or } B)}{p(x)} \quad (5.1)$$

where in this case of two categories,

$$p(x) = \sum_{j=A}^{B} p(x \mid w_j) P(w_j), \text{ where } (w_j \text{ is class}) \quad (5.2)$$

Bayes' formula is expressed informally:

$$posterior = \frac{likelihood \times prior}{evidence} \quad (5.3)$$

3 CURVE FITTING

To understand curve fitting, let's look into the simple example of polynomial curve fitting. Let's say we have a real value input variable x, and we want to predict the real value target variable t. Suppose there is a training data sample of size N observations of x, $\mathbf{x} = (x_1, x_2, \ldots x_N)^T$ with target sample of $\mathbf{t} = (t_1, t_2, \ldots t_N)^T$; these values are

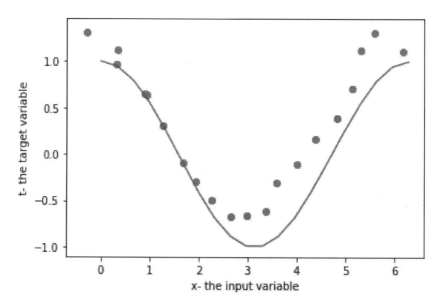

FIGURE 5.2 Training sample plot N = 20 points as red dots, where each dot represents the observation x along with the corresponding target variable t. The curve shows the function $cos 2\pi x$ used to generate the data. We wish to predict the target 't' for given 'x' without perceiving the function.

Pattern Recognition

generated by a function $cos2\pi x$. Figure 5.2 shows the function and training sample plot. Each gray point is the result of adding random noise with Gaussian distribution to the value computed by function $cos2\pi x$ to obtain the target 't'. The goal is to predict the value $\hat{t}$ by giving the new input $\hat{x}$ in the absence of the information of the function caused the points i.e. $cos2\pi x$. Let's assume it a polynomial function y(x, w) where x and w represents the input and coefficients.

$$y(x,w) = w_0 + xw_1 + x^2 w_2 \ldots\ldots\ldots + x^M w_M \tag{5.4}[1]$$

where M is the order of polynomial function.

The coefficient values are computed by fitting the polynomial function to the training data. The fitting process is done by minimizing the error function. The common type error is sum of squared errors, which is the sum of the distances between target and predicted values see Equation (5.5) and Figure 5.3.

$$E(w) = \frac{1}{2}\sum_{n=1}^{N}\{y(x_n,w) - t_n\}^2 \tag{5.5}$$

The goal is to minimize error E(w) value to minimum by substituting the coefficient values to the polynomial equation. Polynomial equations use orders to represent complexity of an equation, differentiation between order M of a polynomial equation is shown in Figure 5.3. The order of polynomial equation represents the complexity of the model, or in other words, the order M is responsible for fitting line properties. In Figure 5.3, we can observe that $M = 0$ and $M = 1$ are a poor fit on data generated by function $cos2\pi x$, but as the polynomial order goes up, the curve fitting improves,

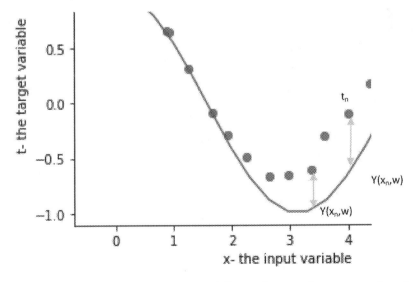

FIGURE 5.3 Sum squared error is the squared distance between the target t_n and predicted value y(x,w).

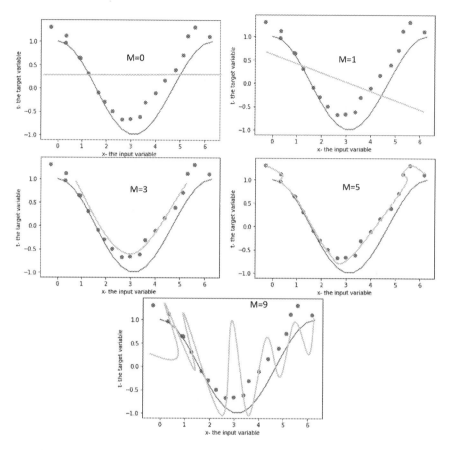

FIGURE 5.4 The polynomial curve fitting with different order (M).

and further increment in order just makes the function more complex and a poor fit. One important thing to observe from Figure 5.4 is with $M = 0$ and $M = 1$, the white line is a poor fit to the data points, and this process is called as underfitting. When the order increases and the line is the perfect fit to the points when $E(w) = 0$, this process is called as overfitting. They both have disadvantages and result in poor performance; see Figure 5.5 to understand the model selection.

Other types of error are RMS (root mean squared error), which allows us to compare different sizes of data sets and ensures scaling with target variable; see Equation (5.6).

$$E_{RMS} = \sqrt{2E(w)/N} \qquad (5.6)$$

3.1 Bias and Variance

Machine learning requires input to train the data. The training data comprises the feature vectors and class labels (see Table 5.1 for an example) or variety, which represents

Pattern Recognition

class labels and sepal.length, sepal.width, petal.length, petal.width represent feature vectors. The trained model is then tested on the data sample which has same feature vectors as the training data. The model predicts the class label based on the learning from the training data. The model predictions are called original value or truth value or target value, and the training set is called as expected value or input value.

Every model works, but which model to select is the real question. The bias and variance trade-off is important in order to select best model.

A bias of an estimator[1] is the difference between this estimator's expected value and the true value of the parameter being estimated (Bias of an Estimator, 2021).

A bias occurs when an algorithm generates results that are systemically influenced due to incorrect assumptions in the machine learning process.

A bias is the difference between the average prediction of model and the expected value (Singh, 2018). Models with high bias are exceedingly simplified and result in high error on training data.

Variance specifies the spread of or variability of the predicted data. High variance results in low error on the training set but high error on test data. Let's look at Figure 5.5 to understand bias and variance.

3.1.1 Bias Variance Trade-Off

If the selected model is oversimplified, it results in underfitting, and if the model is complex, it results in overfitting. The bias variance trade-off represents the right

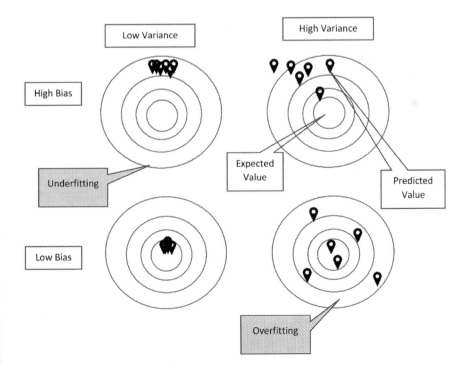

FIGURE 5.5 The bias variance illustration using bull's-eye example.

TABLE 5.1
Sample data for model train/test

Variety/Class	sepal.length	sepal.width	petal.length	petal.width
Setosa	5.1	3.5	1.4	0.2
Setosa	4.9	3	1.4	0.2
Setosa	4.7	3.2	1.3	0.2
Versicolor	7	3.2	4.7	1.4
Versicolor	6.4	3.2	4.5	1.5
Versicolor	6.9	3.1	4.9	1.5
Virginica	6.4	3.1	5.5	1.8
Virginica	6	3	4.8	1.8
Virginica	6.9	3.1	5.4	2.1
Virginica	6.7	3.1	5.6	2.4

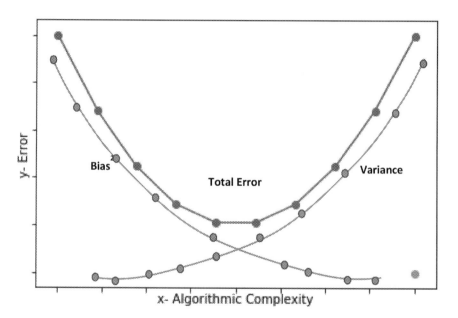

FIGURE 5.6 Bias variance trade-off, where x axis is algorithmic complexity and y axis is error.

balance between these two. Figure 5.6 shows the bias and variance curve along with total error, where the aim is to find the sweet spot with balanced bias and variance. Equation (5.7) represents the relationship between these three.

$$Total\,Error = Bias^2 + Variance + Irreducible\,Error \quad (5.7)$$

4 THE CURSE OF DIMENSIONALITY

Bellman (1961) introduced curse of dimensionality as the number of samples needed to estimate an arbitrary function when a given level of accuracy grows exponentially with respect to the number of input variables (i.e. dimensionality) of the function.

The process grows analytically or computationally complicated with a high number of features, aka high dimensions in comparison to lower dimensions, and these complications are challenging to determine. Let's consider an example in Figure 5.7, which represents two clusters – cluster 1 and cluster 2 – on an x, y axis. If there are two features distinguishing both clusters from each other, the feature values would look like those shown in Table 5.2, i.e. both feature variables are binary. This is a two-dimensional representation of Figure 5.7. If the same set of clusters is represented by a higher number of features or by higher dimensions, it becomes challenging to distinguish each entity in higher dimension, and every data point seems to have similar distances from other data points in multi-dimensional space; see Table 5.3. The model will not be able to learn the data that has an equal number of features and samples. Resolution to the curse of dimensionality is provided by dimensionality reduction algorithms such as PCA (principle component analysis), LDA (linear discriminant analysis), and feature selection.

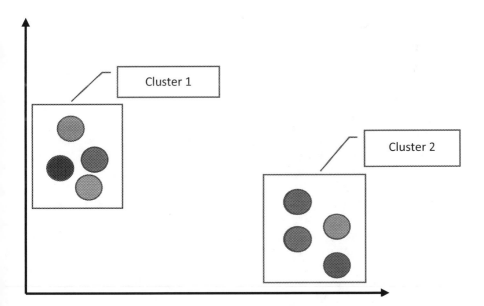

FIGURE 5.7 Illustration of two clusters with similar features in 2D space.

TABLE 5.2
Feature values of clusters shown in Figure 5.7: The two features (two dimensions) are Red and Blue and have binary variables (0 or 1) to represent feature values

Color	Red	Blue
Blue	0	1
Light Blue	0	1
Violet	0	1
Indigo	0	1
Orange	1	0
Pink	1	0
Red	1	0
Maroon	1	0

TABLE 5.3
Representation of two clusters in Figure 5.7 with higher dimensions/features (8 dimensions)

Class	Pink	Maroon	Orange	Red	Blue	Light Blue	Violet	Indigo
Blue	0	0	0	0	1	0	0	0
Light Blue	0	0	0	0	0	1	0	0
Violet	0	0	0	0	0	0	1	0
Indigo	0	0	0	0	0	0	0	1
Orange	0	0	1	0	0	0	0	0
Pink	1	0	0	0	0	0	0	0
Red	0	0	0	1	0	0	0	0
Maroon	0	1	0	0	0	0	0	0

5 CLASSIFICATION

Classification is a predictive modelling technique in which a mathematical model predicts the outcome class given a sample input. Classification classes are also called categories, and the model takes the decision on test data by learning from training data. The classification process is illustrated in Figure 5.8. The process steps start from data collection and follow preprocessing, feature selection, and classification. This process flows bottom-up and top-down after analysing model performance at least after one iteration until convergence.

The example of classification is categorizing a flower in different classes (versicolor, setosa, virginica) based on their features like petal sepal.length, sepal.width, petal.length, petal.width (see Table 5.1). Figure 5.9 illustrates the plot of iris data set (sample in Table 5.1), and the clusters represent different classes. The classification process is required to take a decision based on the decision boundaries, aka curve fitting. Consider the example shown in Figure 5.9 and imagine how a model would differentiate between setosa class and virginica class. It can be done by a straight line between two clusters, which is called linear classification. Consider one more scenario in which the model differentiates between the virginica and versicolor classes. Now if a linear decision boundary performs poor classification, then non-linear decision boundaries are useful (curve, polynomial curve); see Figure 5.10.

Pattern Recognition

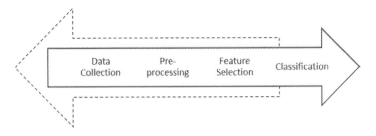

FIGURE 5.8 Classification process steps top-down and bottom-ups.

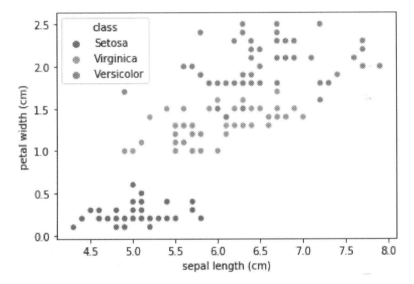

FIGURE 5.9 A plot of iris data set distinguishing between three classes: x axis is sepal length (cm) and y axis is petal width (cm). The three clusters in 2D space represent three classes (setosa, virginica, versicolor).

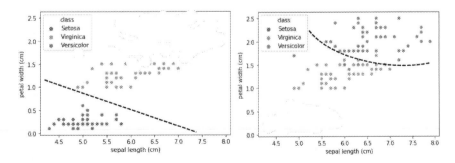

FIGURE 5.10 The classification decision boundary on iris data set shown in a sample in Table 5.1. the left example shows a linear decision boundary between setosa and virginica; the right plot shows a non-linear decision boundary between classes virginica and versicolor. All three classes are illustrated in Figure 5.9.

6 CLUSTERING

Clustering is a process in which categorization is performed by grouping similar items together. Clustering is usually used for unsupervised pattern recognition in the absence of labels/classes. The need for clustering in real life is for identification of similar users or items in a system. For example, it can be used for targeted advertisement, to enhance the user experience by displaying posts by like-minded individuals to each user, to facilitate the creation of social networks in a website, or to identify similar symptoms caused by micro-organisms or viruses.

NOTE

1 An estimator is a rule for calculating an estimate of a given quantity based on observed data.

REFERENCES

Bellman, R. (1961). *Adaptive control processes: A guided tour Princeton University Press.* Princeton, NJ, USA: Princeton University Press.

Bias of an estimator. (2021, May). Retrieved from en.wikipedia.: https://en.wikipedia.org/wiki/Bias_of_an_estimator

Bishop, C. M. (2006). *Pattern recognition and machine learning.* Springer-Verlag New York: Springer.

Dale, A. I. (2005). Thomas Bayes, an essay towards solving a problem in the doctrine of chances (1764). In *Landmark writings in Western mathematics 1640–1940* (pp. 199–207). Elsevier Science.

Liu, Z. P. (2015). Deep learning face attributes in the wild. *Proceedings of the IEEE International Conference on Computer Vision* (pp. 3730–3738).

Singh, S. (2018, May 21). *Understanding the bias-variance tradeoff.* Retrieved from https://towardsdatascience.com/: https://towardsdatascience.com/understanding-the-bias-variance-tradeoff-165e6942b229

6 Automated Pattern Analysis and Curated Sack Count Leveraging Video Analysis on Moving Objects

Ritin Behl, Harsh Khatter, & Prabhat Singh

CONTENTS

1	Introduction	70
2	Problem Statement	70
	2.1 Related Work	71
3	Proposed Approach	72
	3.1 Model Development	73
	Step 1: Problem Definition	73
	Step 2: Gathering Information	73
	Step 3: Training Data	73
	Step 4: Feature Extraction	73
	Step 5: Classification	73
	Step 6: Testing Data	73
4	Proposed Methodology	74
	4.1 System Design	75
	4.1.1 Data Flow Design	75
	4.2 Sequence Diagram	75
5	Implementation Details	75
	5.1 TensorFlow Framework	75
	5.2 Faster R-CNN Algorithm	76
	5.3 Inception Model	78
	5.3.1 Data Collection	79
	5.3.2 Labelling the Data Set	80
	5.3.3 Generating TensorFlow Records for Training	80
	5.3.4 Configuring Training	81
	5.3.5 Training Model	81
5	Conclusion	82
References		82

DOI: 10.1201/9781003169550-6

1 INTRODUCTION

From the advances in car technology to the various features that help drivers in modern cars, the long way of automotive technology is becoming increasingly sophisticated. At the same time, infrastructure transportation providers and resources will increase their reliance on computer vision (CV) to improve their safety and efficiency in transportation. The concept of computers in this way helps to solve critical problems at the transit level, the consumer level and the infrastructure provider level [1].

The demand for limited transportation infrastructure is constantly growing and leading to traffic delays. Accidents and traffic congestion have also had serious economic consequences. Technological advances such as the concept of computers play a key role in solving these types of problems in efficient and effective ways such as traffic charging, incident detection and management, traffic monitoring, traffic monitoring and vehicle control and many more alternatives, with a wide range of applications that computer vision technology can support. Self-help assistance programs are being rolled out at increasing prices, and these programs will begin to shift their role from one aid to another to facilitate decisions as they are related to safety; but the power issue is there with electric automotive to connect the infrastructure [2].

The aim of the independent acquisition programs is to reduce traffic congestion, to produce less expensive cars and emergency equipment, to improve public safety, to reduce environmental impacts, to improve mobility data, and so on. The technology helps regions, cities and towns at the national level to meet the growing demands of a better travel system. The effectiveness of an independent system is largely based on the efficiency and completeness of the acquisition technology [3,4].

Car detection by video cameras is one of the most attractive and inaccessible technologies that requires large data collection and use of vehicle control and management systems. Object discovery is also the basis for tracking. Proper discovery of an object results in better tracking of the object. Modern computer-controlled systems have more complex access requirements than those used by conventional traffic controllers as one road sign is used to represent the multiple devices, these can be many explosive devices [5].

2 PROBLEM STATEMENT

Often it happens human observation is required to keep track of objects unloaded from a truck to the warehouse or loaded from the warehouse into the trucks. There is a possibility of errors. To solve this problem, a web application will be designed which will count the number of sacks that are loaded into the truck or unloaded from the truck. A camera would be used which will be monitoring the sacks that are loaded and unloaded, and it would be easier to find if the sacks loaded are equal to the number of sacks unloaded or not. With the advancement in video analytics, it is feasible to remove the human observation and replace it with a camera-based video analytics solution by object detection method [6–8].

This sack counting system is one of the smart solutions for the transportation industry. There is a chance of error and false results in human observation. The goal of the Sack counting System backend engine is to process the image frames passed

Automated Pattern Analysis

to it via the image processing algorithm according to architectural diagram. The backend engine is trained as such, that, it processes the image frames which was found in the frame folder and predicts the sack count using predictive algorithm and returns live counter of the sacks loaded on to the truck. The engine used is based on the TensorFlow v2 framework, and the image frames are fetched from the live video recording from the CCTV camera recording the video [9,10].

The development of the project will flow in steps one after another such as selecting a model, adapting an existing data set that creates and annotates its own data set, modifying the model configuring the file, training the model and saving the model, moving the model in another piece of software and finally displaying the result of the back-end engine on the custom-developed user interface in the form of a web application.

The data would be collected from a live camera feed, and then the collected video would be converted into frames. The frames will be processed to extract the required information, i.e. the sacks. The model-building algorithm would learn from the frames about the color, shape, size and position of the sacks. Then the model will be tested against some data which is not used for training.

To count the number of sacks a region of interest (ROI) will be determined. The sacks crossing beyond the ROI would be counted, and the counter will be incremented by 1. Initially, the counter would be set to 0. The counter will be refreshed every time the model is run again after terminating the model once. The result output is the command prompt screen of the sack count system containing the number of sacks either loaded or unloaded from the truck. Also, the annotated frames with confidence score are displayed in the output window [11].

2.1 Related Work

Object acquisition is a common term used for CV techniques that classify and locate objects in a photograph. The discovery of the modern object depends on the use of the CNN network. There are many programs suitable for Faster R-CNN, R-FCN, Multibox Single Shot Detector (SSD) and YOLO (You Only Look Once). The original R-CNN methods work with neural net classifier performance on plant sample images using external suggestions (samples are planted with external box suggestions, with the extraction feature performed on all planted samples). This method is very expensive due to the many crops. The Faster R-CNN is able to reduce the scale by doing the feature selection once only in the whole image, and uses the crop on the lower parts of the sample [4,5].

Faster R-CNN goes a step further, using extruded materials to create agnostic box suggestions, i.e. a single-release feature for the whole image, no external box suggestions. R-FCN is similar to Faster R-CNN, but feature harvesting is done on different types of enhancement efficiency. The advantage of the simplicity is that the YOLO model is faster compared to the Faster R-CNN and SSD, which reads representation. This increases the error rate of the area, and YOLO does not work well with images with a new feature rating but reduces the false rating. This will speed up real-time applications. This comes with the price of reduced accuracy [6–8].

SSD with MobileNet refers to the model in which the SSD is the type of SoftNet installer. Accurate trade-off speed and many other modern systems are used for finding something that require real speed. Methods like the YOLO and SSD work faster. This often comes with a decrease in predictability, while Faster R-CNN models achieve higher accuracy but are more expensive to run. The cost per model speed depends on the application. SSD works better, if compared to more expensive models like Faster R-CNN, in smaller images its performance is much lower [11].

3 PROPOSED APPROACH

The objective is to construct a detection and counting model such that it will increase the efficiency of the loading/unloading process in trucks and also increment the efficiency of counting over human observation and lessen the chances of human forgery in the transportation industry. But before constructing the model, we need to study the present scenario and understand the factors or parameters that affect the statistics. The objective was pursued as given in Figure 6.1.

To achieve the objective, the focus was put on data collection and analysis and model research. The detection model or implementation part is also a goal. The parameters and their prioritized values are decided only after proper analysis. For this, a data set as large as possible is required. The discovery of the TensorFlow framework is used to build an in-depth learning network that helps solve object acquisition problems. This includes a collection of pre-defined retrained models trained in the COCO database as well as the Open Pictures Dataset. These types can be used humbly if there is interest in categories on this data only. They are also useful for implementing models when trained in novel data.

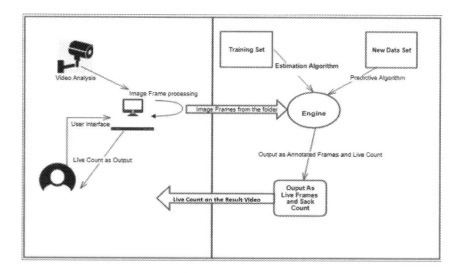

FIGURE 6.1 Architecture diagram of sack count system.

3.1 Model Development

Since the proposed model is based on custom object detection, there will be basic six steps to achieve the goal.

Step 1: Problem Definition

This is the first step in finding an item. Defining a problem carefully requires an understanding of how things will be used, who needs the items and how the adoption function fits into an organization that needs to predict. Job description, outsourcing, decision-making of a project, general discussion of trade-offs (accuracy vs. speed) and setting up a project codebase are the basic and fundamental requirements of a particular problem.

Step 2: Gathering Information

There are at least two types of information required: (a) image data and (b) live camera feeds. Usually, it will be difficult to get enough live video data to be able to fit a good image capture model due to the large amount of background noise. However, sometimes, older data will not be as useful due to changes in the upload method. For example, today, conveyor belts are more popular than handicrafts.

Step 3: Training Data

Always start by labeling details. Training data are images collected as samples and interpreted to train deep neural networks. With the discovery of something, there are a lot of methods used to fix and quote data for training purposes. Many popular ways to distribute databases such as Pascal containing LabelImg tools.

Step 4: Feature Extraction

Feature extraction is a basic step used by automated methods based on machine learning methods. Its main purpose is to extract useful features or important components from the data, which the computer can detect to calculate values in input images. An element is defined as the work of one or more dimensions that describe a particular asset (such as color, texture) of the whole image or layer or object.

Step 5: Classification

This feature provides a collection of records seen by a group of features x and a class label y. It is aimed at describing a class model that includes a recording class label. Image classification follows steps such as preprocessing, component fragmentation, feature extraction and segmentation.

Step 6: Testing Data

There is test data within the file which holds up to 30% of the full data set. This test data isn't being employed for training the model. Even after training the model, it is required to check the model on some data set. It's very significant to understand the accuracy of the full model and the confidence score of a selected test run on a trained model.

4 PROPOSED METHODOLOGY

The goal of the sack counting system is to process the image frames passed to it via the image processing algorithm mentioned in Figure 6.2. The engine is trained such that it processes the image frames found in the frame folder and predicts the sack using a predictive algorithm and returns the live count of the sacks loaded on to the truck. The engine used was the OpenCV, and the image frames are fetched from the live recording from the CCTV camera recording the video. The external interface of this engine begins with the collection of frames saved in a .csv folder.

The frames are extracted through a live video stream of cameras installed at the work area. Then the OpenCV engine takes the frames one by one from the folder and starts analyzing them through a command-line interface. The engine then identifies the correct frames and increments the counter; now the frames become annotated. Finally, the counter frames, annotated frames and the desired result will be the output video. The annotated result video will become the output. The internal interface will be decomposed into two parts: Train and Test.

The Train part will be used to train the model for detecting the frame containing a sack. Initially, data set is required and divided randomly into training and validation sets. It contains m image frames along with the label "sack" marked on that frame using LabelImg annotation. Here, the engine will call the main function, which in turn calls the train function, which first initializes the parameters randomly and updates weights at each iteration up to some minimum cost. The model hyperparameters according to the variance and bias provided by the model are tuned. It is a continuous process to make a model robust to new images and provide more accuracy. This process will train the model, and final parameters can be used to predict the new frame.

In the test part or operational phase, the main function will be called, which refers to the predict function. The predict function will check the image frame according to the trained model parameters. If the sack is present in an image frame, then the result would be there, in the form of an output result video which will show the number of sacks loaded/unloaded into truck and time and date of the loading/unloading in the truck.

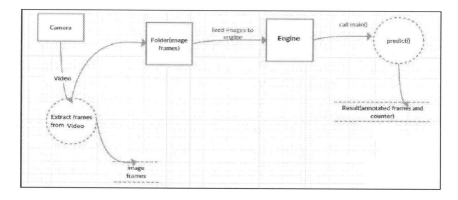

FIGURE 6.2 Data flow diagram of the proposed method.

Automated Pattern Analysis

4.1 System Design

4.1.1 Data Flow Design

The data flow diagram (DFD) shows the flow of information for any process or system. Data flow diagrams represent systems and processes that will be difficult to define in textual format.

4.2 Sequence Diagram

A sequence diagram shows the object interactions that are arranged in time sequence. It would describe the classes and objects which are involved in the sequence of messages and exchange between the objects needed to carry out the functionality of the structure.

5 IMPLEMENTATION DETAILS

5.1 TensorFlow Framework

TensorFlow is an open-source framework used to integrate and build great machine learning programs. It was created by the Google Brain team. It incorporates mechanical and in-depth learning methods as well as neural network models and algorithms and makes them useful in the form of a standard platform. It uses Python as a programming language that provides a simpler and better complete API for building applications using a framework and also aids in the development of those applications in greater C ++ functionality. TensorFlow helps with prediction of production at a certain level, with the same models used to train a deep neural network. It can train and operate deep neural networks – for example, neural networks used for image recognition, handwriting digitization, duplicate neural networks, embedding words, sequential machine translation models etc.

Python is supported by TensorFlow to provide all framework functions. Python is a simple structured language. It is easy to read and work in and offers useful options

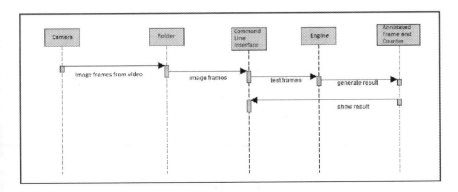

FIGURE 6.3 Sequence diagram of the proposed approach.

for explaining how high-end manufacturers can be integrated. Nodes and tensors are Python objects, and the provided Python applications are TensorFlow applications themselves.

In Python, real mathematical functions are not performed. C ++ phones that are known for high-performance work with conversion libraries available for language processing, as well as PDE (which is part of the separate classification) are used.

TensorFlow allows developers to create dataflow graphs. Dataflow graphs are defined as structures that show how data moves across a graph, or processing nodes in a series. Mathematical performance is represented by nodes in the graph. Each edge between the areas is a multi-distance data system called tensor.le in the TensorFlow framework. API and packages are provided by Python, which simply directs traffic between pieces and connects loopholes at intersections.

TensorFlow applications can be distributed to any target and are user friendly, be it machine, cloud collection, iOS and Android devices CPUs or GPUs. If you use Google's own cloud, you'll be able to use TensorFlow in the Google Silicon TensorFlow Processing Unit (TPU) to keep up the speed.

Output models of the TensorFlow framework, however, will be used on any device that can be used to provide prediction. TensorFlow 2.0 analyzes the framework in some ways that support user feedback to make it easier to find that (e.g. by using the simple Keras API for training) and to do more. Distributed training is very complex to run due to the replacement API, and support for TensorFlow Lite enables us to deploy models in a large platform style. However, the code written for previous versions of TensorFlow must be rewritten only occasionally, and sometimes significantly, so that the maximum benefit of the latest TensorFlow 2.0 features can be established.

One of the great advantages of TensorFlow is that it offers the advantages of machine learning development. Instead of managing the nitty-gritty details of using algorithms or determining the appropriate ways to capture the first result of a task in another installation, the developer can work specifically on the general concept of the application. TensorFlow takes care of the small print behind the scenes.

Also, TensorFlow offers one more benefit for developers who need to fix a problem and take TensorFlow details. In the framework, there is a mode for frequent use that allows one to test and modify the performance of each graph especially, and without stitching, instead of naming the graph as one opaque object and testing it all at once. And the TensorBoard Visualization Suite allows one to view and select a graph-driven interactive dashboard.

TensorFlow also received a chance for Google support. Google has not only allowed for faster growth after the project but has created many important TensorFlow methods that make it much easier to use and operate, for example, the TPU silicon for faster performance in the Google cloud, your browser and the desired size for the frame, the hub for online frame sharing and much more.

5.2 Faster R-CNN Algorithm

R-CNN was launched in 2014 and has attracted a lot of interest in the computer vision community. The whole idea behind R-CNN is to implement a search engine selected to propose 2,000 region-and-interest (ROI), which is then fed into the convolutional

Automated Pattern Analysis

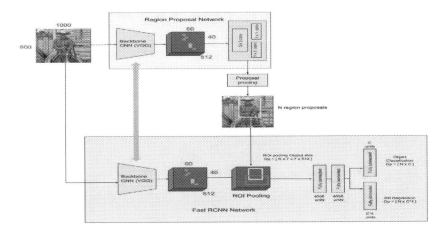

FIGURE 6.4 Faster R-CNN algorithm.

neural network (CNN) to collect features. These features have been used to classify photographs and their object boundaries using SVM (support vector machine) and regression methods. For a more detailed explanation, see this section. And this approach was quickly followed by R-CNN, which became a faster and better way to identify an object. Faster R-CNN uses an ROI pooling system that shares features throughout the image and uses a modified type of spatial pyramid pooling method to efficiently capture features over time. In the case of Faster R-CNN, it is still slow because it has SS to handle, which is computationally very slow. The time to test Faster R-CNN takes from 47 seconds to 0.32 seconds, and 2 seconds to generate 2,000 ROIs. It adds up to 2.3 seconds per image.

The shortcomings of the two algorithms paved the way for researchers to quickly come up with R-CNN, in which the test time for each image with field resolutions was only 0.2 seconds. This is due to the latest approach given the completely different model used for end-to-end training.

There are two proposals for Faster R-CNN, such as Region Proposal Networks (RPNs) used to generate field proposals and networks that use these found resolutions to identify objects. The big difference here with Faster R-CNN is that the latter fields use selective search to obtain proposals. The time and cost of building field resolutions is much lower in RPNs than in selected searches, as RPNs share very important calculations with the detection network. However, RPNs call field boxes anchors and propose them as objects.

The Faster R-CNN Object Detection Network feature is designed with an extraction network that can often use a blocked traditional neural network. After this, two subnetworks can be trained. The primary may be the RPN, which is accustomed to generating object conclusions. The latter is therefore used to estimate the specific class of a given article. The primary variant for Faster R-CNN is RPN, which is added after the final composite layer. It is often trained to provide field conclusions without the need for external mechanisms such as selective search. ROI pooling can be used as an upstream classification and bounding box resistor similar to the one used in Faster R-CNN.

5.3 INCEPTION MODEL

The beginning may be the deeply controversial neural specification introduced in 2014. It won the ImageNet Large-Scale Visual Recognition Challenge (ILSVRC14). It was mostly developed by Google researchers.

In convolutional neural networks (CNNs), the outer part of the work is performed at the appropriate layer for use between the most common options (1 × 1 filter, 3 × 3 filter, 5 × 5 filter or maximum pooling). We all want to find the right local structure and replicate it spatially.

As each of these inception modules, is stacked on top, their output correlation statistics vary: as the properties of the upper extraction are captured by the higher layers, their spatial concentration of 3 × 3 and k reduction is estimated, indicating the ratio; 5 × 5 resolutions increase as we move to higher layers.

However, the computational cost of such an answer would increase enormously. For this reason, within the figure, diffusion reduction is used as diffusion reduction methods by 1 × 1 convolution.

The installation module aims to act as a "multi-level feature extractor" by calculating the 1 × 1, 3 × 3 and 5 × 5 additives in the homogeneous modules of the network – the output of those filters being then stacked together. Channel size and before feeding in the bottom layer of the network.

The original version of this architecture was called GoogleNet, but later versions were called Inception VN, where N refers to the version number entered by Google. The Inception V3 architecture included in the Kears Core comes from a later publication by Szegedi. Inception Architecture for Computer Vision (2015) is proposing an update of the Inception module to further increase image net classification accuracy. Initially, the V3 weighed 93 MB, which is smaller than both VGG and ResNet.

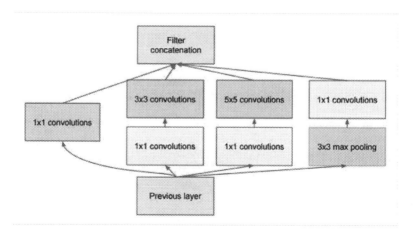

FIGURE 6.5 Inception model.

Automated Pattern Analysis 79

5.3.1 Data Collection

The classifier is built in a taxonomy to identify sacks. the data set is a very important thing in creating taxonomy. This may be the basis for our classification, on which object detection ceases. Various and varied images containing objects are collected. Then the directory name image inside the research directory is created. Now, 80% of the photos are stored in the train directory and 20% of the images are stored in the test directory inside the Pictures directory. We have collected 174 images in the train directory and 35 images in the test directory. The results are shown in Figures 6.6 and 6.7.

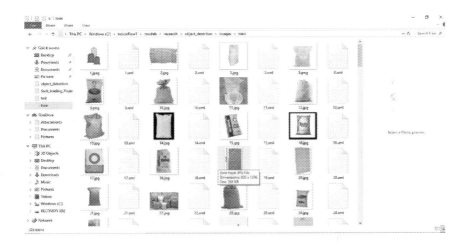

FIGURE 6.6 Train directory.

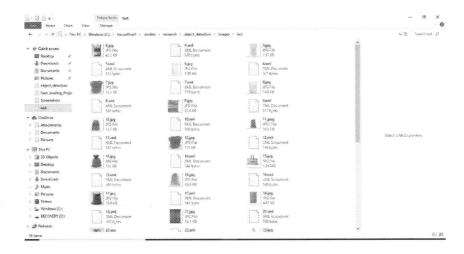

FIGURE 6.7 Test directory.

5.3.2 Labelling the Data Set

Open the LabelImg tool and start drawing rectangular boxes on the image where the subject is. Label them with the appropriate name as shown in Figure 6.8. Save each image created by labeling an xml file with the corresponding image name as shown in Figure 6.9.

5.3.3 Generating TensorFlow Records for Training

For this step, we want to create TFRecords that can be displayed as an input file to train the subject detector. Xml_to_csv.py and Gener_tfrecord.py codes are given. Create test.csv and train.csv files in the xml_cs_csv.py images folder. Then create

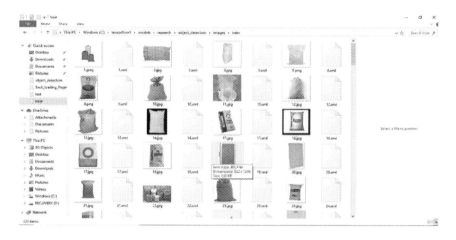

FIGURE 6.8 Train directory with xml file.

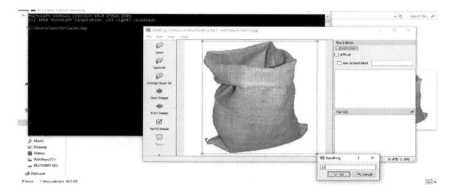

FIGURE 6.9 LabelImg tool.

Automated Pattern Analysis

files for recording in TensorFlow by executing the appropriate commands from the Object_Detection folder.

5.3.4 Configuring Training

Create a replacement directory called Training in the Object_Detection directory. Use the text editor to create the restore file and place it as a label map in the training directory. The label tells the map instructor what each object is by defining the mapping of the advanced name to the class ID number. Now, add the content to classmap.pbtxt then switch to the format to create a label map for your taxonomy. The label map ID number should be as defined in the Generate_tfrecord.py file. We need a model algorithm to train our classification. During this project, we are using the faster_rcn_inception model. TensorFlow's Object Detection API includes a large number of models. Now, open the file using the text editor and make the necessary changes to the fast_rcnn_inception_v2_pets.config file saved in the directory.

5.3.5 Training Model

At the situation object_detection/legacy/ find the file train.py. Open the object detection directory and copy the train.py file and paste it within the same directory. Run the subsequent command to begin training the model in the object detection folder itself. It takes around one to two minutes to start the setup before the training begins. The training starts, and it looks like Figure 6.10.

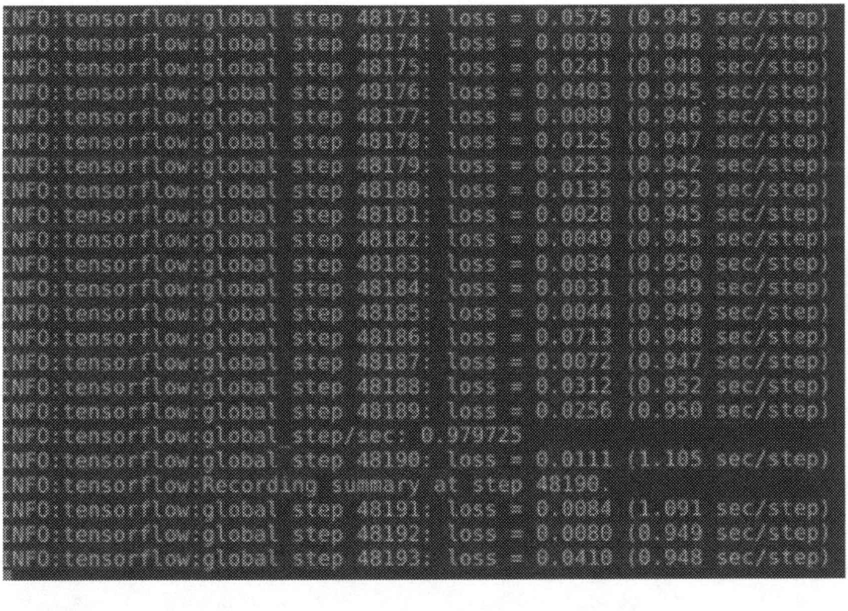

FIGURE 6.10 Training model.

5 CONCLUSION

The work done so far is concentrated on the collection of image data and labelling from and training the model for object detection in an image. Also, the focus was on different algorithms that are there and their comparative study for analysis of video data. The detection model will take these resulting parameters as input and produce a prediction. The goal to convert this analysis into an accurate detection and counting using these object detection and tracking algorithms has been achieved.

The framework used in this project for the application of object count, the TensorFlow framework, can be further extended and optimized through various means. It can make simple but significant modifications in the model while training and during data preprocessing like increasing the training size, more epochs in training and bigger images for improving the accuracy. Also, we can implement this whole idea on mobile devices using MobileNet (included in TensorFlow framework), which provides better accuracy.

REFERENCES

[1] Huang, J., V. Rathod, C. Sun, M. Zhu, A. Korattikara, A. Fathi, I. Fischer, Z. Wojna, Y. Song, S. Guadarrama, "Speed/accuracy trade-offs for modern convolutional object detectors", arXiv preprint arXiv:1611.10012.

[2] Redmon, J., S. Divvala, R. Girshick, A. Farhadi, "You only look once: Unified, real-time object detection", Proceedings of the IEEE Conference on Computer Vision and Pattern Recognition, Las Vegas, NV, USA, 2016, pp. 779–788.

[3] Liu, W., D. Anguelov, D. Erhan, C. Szegedy, S. Reed, C. Fu, A. C. Berg, "SSD: Single shot multibox detector", European Conference on Computer Vision, Springer, Amsterdam, The Netherlands, 2016, pp. 21–37.

[4] Galitsky, Boris, "A content management system for chatbots", in Developing enterprise chatbots, Vol. 1, Springer, Nature Switzerland, pp. 253–326, April 2019.

[5] Ali, Nawaf, Musa Hindi, Roman V. Yampolskiy, "Evaluation of authorship attribution software on a chatbot corpus", International Symposium on Information, Communication and Automation Technologies, Taipei, Taiwan, October 2011.

[6] Go, Eun, S. Shyam Sundar, "Humanizing chatbots: The effects of visual, identity and conversational cues on humanness perceptions", in Computers in Human Behavior, Vol. 97 (2019), pp. 304–316.

[7] Singh, P., H. Khatter, S. Kumar, "Evolution of software-defined networking foundations", in Evolution of Software-Defined Networking Foundations for IoT and 5G Mobile Networks, IGI Global, USA, Vol. 1, Pp. 98–112, 2021.

[8] Khatter, H., B. M. Kalra, "A new approach to Blog information searching and curating", in Proceedings to Sixth International Conference on Software Engineering CONSEG, Indore, India, 2012, pp. 1–6.

[9] Khatter, H., A. K. Ahlawat, "Analysis of content curation algorithms on personalized web searching", in Proceedings of the International Conference on Innovative Computing & Communications (ICICC), New Delhi, 2020, pp. 1–4. http://dx.doi.org/10.2139/ssrn.3563374

[10] Khatter, H., M. C. Trivedi, B. M. Kalra, "An implementation of intelligent searching and curating technique on blog web 2.0 tool", in International Journal of u -and e-Service, Science and Technology, Vol. 8, No. 6 (2015), pp. 45–54.

[11] Khatter, H., A. K. Ahlawat, "An intelligent personalized web blog searching technique using fuzzy-based feedback recurrent neural network", in Soft Computing, Vol. 24, No. 12 (2020), pp. 9321–9333. doi: 10.1007/s00500-020-04891-y

7 DBSU
A New Fusion Algorithm for Clustering of Diabetic Retinopathy Disease

Sanjay Kumar Dubey, Tanvi Anand, & Rekha Pal

CONTENTS

1	Introduction	83
2	Related Work	84
3	Research Methodology	86
	Data Set	87
	K-Means Clustering	87
	DBSCAN Clustering	88
	DB-SU (Proposed Work)	89
4	Experimental Results and Performance Evaluation	90
	Evaluation Performance through Accuracy	90
	Evaluation Performance on the Basis of Error Rate	91
	Evaluation Performance on the Basis of Precision Recall	92
	Evaluation Performance through F-Measure	93
5	Conclusion and Future Scope	94
References		95

1 INTRODUCTION

Diabetic retinopathy is one of the serious retinal disorders because of the diabetes that leads to the loss of vision. It is a disease which increasing day by day. According to the World Health Organization (Diabetes, WHO, 2021), approximately 422 million people are facing the problem of diabetes. It is one of the leading causes of death worldwide. Diabetic retinopathy may not have symptoms at the early stage of the disease, and it is commonly found in patients between the ages of 30 and 69. Therefore, it is recommended to have regular eye check-ups, but due to the lack of specialized oculists, this regular check-up involves higher cost. To resolve this, a low-cost computer-aided device can be used to provide advance diagnosis and treatment. People suffering from type 1 diabetes don't suffer from diabetic retinopathy for approximately the first five years after diagnosis, and in type 2, one-fifth of people are suffering from diabetic retinopathy at the time of diagnosis. However, after fifteen years, almost all the people with type 1 and type 2 diabetes are found to be

DOI: 10.1201/9781003169550-7

suffering from background retinopathy. According to the oculist, micro aneurysm is the microscopic filling of blood in artery walls and hemorrhages which cause damage to the tissue of back wall of the eye and is visualize as red spots slightly larger than micro-aneurysm, most important signs that ensure that patient is suffering from diabetic retinopathy (Saranya & Kunthavai, 2015). Both hard and soft exudates play a very important role in tracking the progress of treatment and grading it. Exudate detection is more preferable if done by computer-aided devices and provides more precise and fast results for diagnosis and also helps the oculist to treat it accordingly (Rajput & Patil, 2014).

Data mining plays an important role in the health care industry in analyzing and dealing with large amounts of data. Hence, with data mining, better diagnosing and treatment of clinical problems are achieved. Health care is bound with large amounts of information in unstructured form about patients, diseases, medical equipment, treatments, etc., that must be processed and analyzed for knowledge retrieval.

The clustering technique is a very important part of data mining that doesn't have a predefined class; it combines that data into clusters depending upon similarity. Its unsupervised techniques – include DBSCAN, K-means, hierarchical, etc. – are beneficial in health care in diagnosing many diseases like ailments of the heart, lungs, liver, etc. Disease detection has become crucial due to the increase in new disease. Proper diagnosing and treatment can then be made available early on to prevent progression to the severe stage.

Diabetic retinopathy is a very dangerous disease. If it is not detected at the early stage, it leads to blindness. Therefore, an accurate algorithm is required for investigating the retinal blood vessel features, which helps the oculist and eye care specialist to use this feature of retinal blood vessels for early detection of the disease.

The main aim of this chapter is to propose a new hybrid algorithm, DB-SU form, by merging two algorithms, DBSCAN and SURF, for combining essential features to have more rigorous diagnosing of diabetic retinopathy with the extraction of the unique key points along with removal of redundancy and then comparing its performance with the existing DBSCAN and K-means in terms of various parameters.

The remaining part of the chapter is organized as follows: Section 2 includes related work, Section 3 presents methodology used for carrying research, and Section 4 includes experimental results, followed by the conclusion in Section 5.

2 RELATED WORK

The health care industry produces voluminous amounts of data, which provides hidden and interesting patterns that are valuable for decision-making, which leads to many benefits like information about the patient, disease detection with its diagnosis, fraud identification in health insurance, and detection of effective treatment. Data mining techniques in health care help medical practitioners to analyze the data from various dimensions and categories. Data mining techniques comprise classification, clustering, and association rule mining which make the decision-making process simpler for disease detection and treatment.

Classification is a technique which predicts data into predefined classes on the basis of similarity. It is a supervised algorithm which includes SVM, K-NN, DT, Naïve Bayes etc. (Jabbar et al., 2013; De Carvalho et al., 2013; Sindhumol et al., 2013; Shajahan & Sudha, 2014). Association rule mining records the repetition of items in a data set (Ilayaraja & Meyyappan, 2013). Researchers (Arenas-Cavalli et al., 2015) proposed a web-based platform to automate retinopathy screening for diabetes. Clustering is a technique in which no predefined classes exist, but clusters are formed on the basis of similarity, and so it is unsupervised clustering, which comprises K-means, hierarchical, DBSCAN, EM, etc. (Bhuvaneswari et al., 2014). A research study was conducted on a diabetic retinopathy barometer towards retinopathy screening and treatment (Cavan et al., 2017). A group of researchers (Mwangi et al., 2018) proposed the effectiveness to increase the awareness for diabetic retinopathy. Research work (Mwangi et al., 2018) focused on adapting guidelines for clinical practices of diabetic retinopathy in Kenya. A team of researchers (Lin et al., 2019) did a survey for diabetic retinopathy screening to perform assessment and verify the quality of the retina-based images. Research work (Long et al., 2019) deployed SVM classification and dynamic threshold to detect hard exudates in color retinal images automatically. The researchers predicted diabetic retinopathy using some data mining techniques (Anand et al., 2019). The feasibility of deep learning techniques in order to predict retinopathy progression in diabetic patients is discussed by the research study from Arcadu et al. (2019). An effective method for human retina optic disc segmentation was proposed with fuzzy clustering methods (Abdullah et al., 2020). Another research team (Mwangi et al., 2020) focused on the feasibility of cluster-controlled effectiveness to increase knowledge on the diabetic-based retinal disease. A research study was performed on the risk factors, with diabetic complications in new cluster-based diabetes subgroups (Tanabe et al., 2020).

Figure 7.1 shows the usage of data mining techniques in a diabetic retinopathy data set and other disease, and from the analysis and research, it clearly shows that the clustering technique is more used in handling diabetic retinopathy data as compared to other techniques. Hence, the clustering technique is chosen for handling diabetic retinopathy.

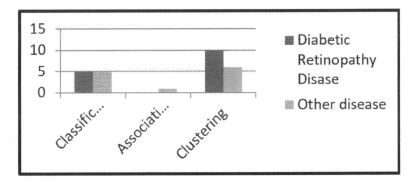

FIGURE 7.1 Data mining techniques used in diabetic retinopathy and other diseases.

Figure 7.2 clearly shows that diabetic retinopathy is increased across the world, and fuzzy c-means techniques are widely used for analysis, while K-means and DBSCAN are used much less. Hence, K-means and DBSCAN are used to analyze the performance of handling the large data set.

3 RESEARCH METHODOLOGY

The methodology includes loading of diabetic retinopathy data from the UCI repository, and then it is undergone in the preprocessing stage, which will involve cleaning of data, and then attributes are selected, which will then undergone into K-means, DBSCAN, and proposed DB-SU, which is shown in Figure 7.3.

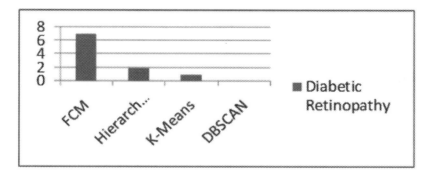

FIGURE 7.2 Diabetic retinopathy analyses using clustering technique.

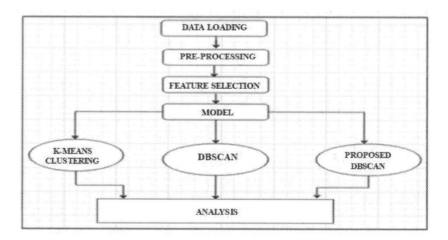

FIGURE 7.3 Research methodology.

Data Set

In this chapter, diabetic retinopathy data is collected from the UCI repository. The data is then passed to the preprocessing stage, in which the data is corrected and null entries are removed. After that, the data is applied in MATLAB. The data set has twenty attributes, and all the values in the data are numeric. The data set has two classes named diabetic retinopathy and non-diabetic retinopathy or healthy person. In the data of 1,151 records, 53% belong to people suffering diabetic retinopathy and 47% belong to healthy persons, as shown in Figure 7.4.

K-Means Clustering

K-means clustering is unsupervised clustering in which the algorithm partitions n data items into k number of clusters in which each data point belongs to one cluster on the basis of the Euclidean distance calculated and the distance to which the data item points, is near to the cluster. (Tapas & David, 2002). The cluster formation is shown in Figure 7.5.

Steps Involved in the Algorithm

1. Initialize the value of k that is the initial centroid of the clusters.
2. Calculate the distance that is the Euclidean distance of each object with respect to the clusters taken initially.
3. Then set the object position in the cluster according to its nearest distance evaluated with the use of the distance formula.
4. Then again evaluate the centroids of the cluster by taking the average of the value of the object.
5. Repeat steps 2–4 until no further change will be obtained in the new cluster values.

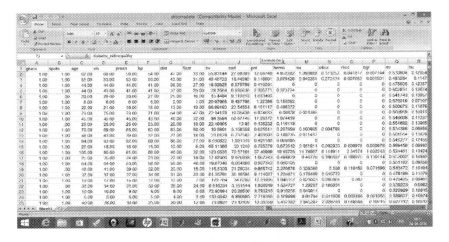

FIGURE 7.4 Diabetic retinopathy database.

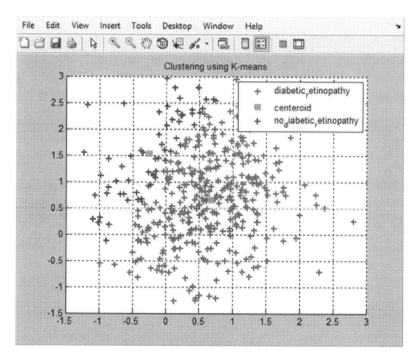

FIGURE 7.5 Clusters formation using K-means.

DBSCAN Clustering

DBSCAN is one of the important density-based spatial clustering algorithms which is used to identify the dense regions which have a maximum number of points close and identify the portion of noise which marks as outliers the low-density regions. This algorithm is based on two important parameters: eps, which termed epsilon, and minpts, which is termed minimum points. Epsilon means the radius around the eps neighborhood, and the minimum points means the points in that radius. The cluster formation is shown in Figure 7.6.

Steps Involved in the Algorithm

1. Initially, consider that every object or data point present in the data set is termed unassigned.
2. Then select a random unassigned data point p within the database depending upon the minimum points and epsilon chosen.
3. The data point will then assign to a new cluster.
4. Then the algorithm will check whether then core point is p or not. If it is not p, then it is considered a noise point, and DBSCAN moves to another unassigned point in the data set.
5. Repeat the algorithm until we reach an assigned point.

DB-SU (Proposed Work)

DB-SU is a hybrid algorithm form from the merging of features of DBSCAN and SURF. DB-SU stands for density-based spatial clustering and application with noise, along with a sped-up robust feature-extraction algorithm for extracting some unique key points and descriptors with respect to class for target attributes of the data set. A set of SURF key points and descriptors can be extracted from a unique 1ns and 0rs value and then used later to detect the same class. The cluster formation is shown in Figure 7.7.

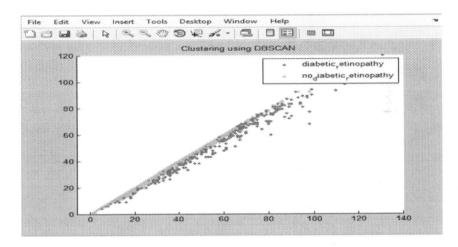

FIGURE 7.6 Cluster formation using DBSCAN.

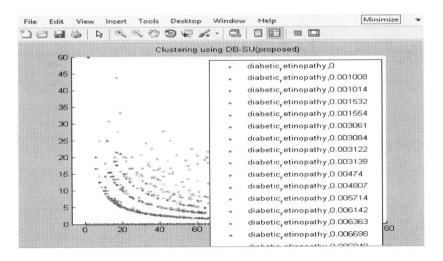

FIGURE 7.7 Clustering using DB-SU (proposed).

Steps Involved in the Algorithm

1. Initially, consider that every object or data point present in the data set is termed unassigned.
2. Then select a random unassigned data point *p* within the database depending upon the minimum points and epsilon chosen.
3. Then fetch the unique key points on the basis of the same minpts value.
4. Assign the same key points a data point with a new cluster.
5. Precise main data point which is not *p* and is also not similar to any data point that is plotted.
6. Then move to the next unassigned data points.
7. Repeat the algorithm until we reach assigned points.

4 EXPERIMENTAL RESULTS AND PERFORMANCE EVALUATION

To visualize the performance of the unsupervised algorithm that are K-means, DBSCAN, and DB-SU (proposed) on diabetic retinopathy data set, several parameters are considered: accuracy, error rate, precision, recall, F-measure for comparing the performance, and validating the work.

Evaluation Performance through Accuracy

The performance of the K-means, DBSCAN, and DB-SU algorithms is evaluated using accuracy, which is calculated by using true positive (TP), true negative (TN), false positive (FP), and false negative (FN).

$$ACCURACY = (TP + TN) / (TP + TN + FP + FN) \qquad (7.1)$$

Figure 7.8 depicts the performance of the K-means, DBSCAN, and DB-SU clustering algorithms for varying iterations. From the graph, it is clearly seen that the accuracy of the proposed technique (DBSCAN-SURF) is better than DBSCAN and K-means. The graph presented the accuracy which is obtained after running the

TABLE 7.1
Accuracy of K-means, DBSCAN, and DB-SU on varying iterations

Number of Iterations	Accuracy of K-Means	Accuracy of DBSCAN	Accuracy of DB-SU
1	5	40	45
3	5	25	30
5	4	15	20
7	2	17	23
10	1	1	10

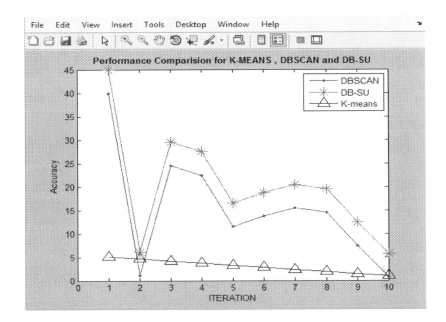

FIGURE 7.8 Accuracy of K-means, DBSCAN, and DB-SU based on varying iteration.

TABLE 7.2
Error rate of K-means, DBSCAN, and DB-SU for varying iterations

Number of Iterations	Error Rate of K-Means	Error Rate of DBSCAN	Error Rate of DB-SU
3	15	6	6
4	14	7	7
6	11	7	7
7	11	8	7
8	10	9	8

DBSCAN-SURF algorithm code for the first time is high and the accuracy is changing with each iteration with the change in the data input in the algorithm.

EVALUATION PERFORMANCE ON THE BASIS OF ERROR RATE

Performance evaluation of K-means, DBSCAN, and DB-SU on the basis of error rate for varying iterations, which is obtained by using true positive rate (TP), false positive rate (FP), false negative rate (FP), true negative rate (TN) as shown in the formula error rate in Equation 7.2.

$$\text{ERROR RATE} = 1 - [(TP+TN)/(TP+TN+FP+FN)] \qquad (7.2)$$

Figure 7.9 depicts the error rate based comparison of K-means, DBSCAN, and DB-SU. The figure clearly depicts that the error rate of DB-SU (proposed) is less than DBSCAN and K-means. The figure shows that the DB-SU is better than K-means and DBSCAN in terms of error rate, and it also shows that the error rate remains constant after four instances of code execution.

Evaluation Performance on the Basis of Precision Recall

Performance of K-means, DBSCAN, and DB-SU clustering algorithm is analyzed on the basis of precision and recall, which depends on true positive value (TP), false positive value (FP), true negative value (TN), and false positive value (FP).

$$\text{PRECISION} = TP / (TP + FP) \tag{7.3}$$

$$\text{RECALL} = TP / (TP + FN) \tag{7.4}$$

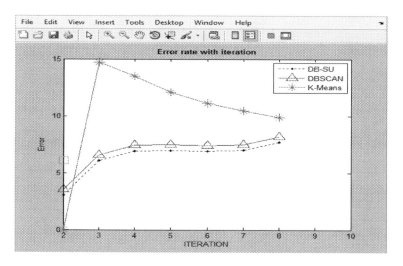

FIGURE 7.9 Error rate of K-means, DBSCAN, and DB-SU on varying iterations.

TABLE 7.3
Precision recall evaluation of K-means, DBSCAN, and DB-SU

Recall	Precision of K-Means	Precision of DBSCAN	Precision of DB-SU
0	1.0	0	0
20	1.0	0	0.4
40	0.8	0.1	0.8
60	0.7	0.3	0.9
80	0.6	0.65	1.0

DBSU

Figure 7.10 shows precision recall analysis of K-means, DBSCAN, and DB-SU. The graph clearly shows DB-SU precision is improving as recall increases, and it is said that precision measures to what extent positive prediction is correct, and recall measures the amount of positive events predicted that is correct. So DB-SU results are more precise as compared to DBSCAN and K-means.

Evaluation Performance through F-Measure

Performance of the unsupervised clustering algorithm is measured by F-measure, which is a harmonic combination of precision and recall, which is evaluated by true positive value (TP), false positive value (FP), and false negative value (FN).

$$F-MEASURE = (2*TP)/(2*TP+FP+FN) \qquad (7.5)$$

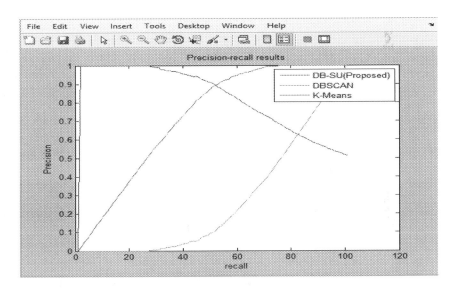

FIGURE 7.10 Precision recall–based evaluation of K-means, DBSCAN, and DB-SU.

TABLE 7.4
F-measure vs. recall result of K-means, DBSCAN, and DB-SU

Recall	F-Measure of K-Means	F-Measure of DBSCAN	F-Measure of DB-SU
0	0	0	0
20	0	1	1.2
40	0.2	1	1.6
60	0.25	1	1.8
80	0.6	1.2	1.7

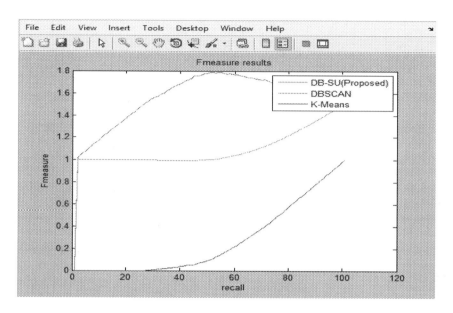

FIGURE 7.11 F-measure-vs.-recall–based evaluation of K-means, DBSCAN, and DB-SU.

Figure 7.11 clearly signifies the performance evaluation of K-means, DBSCAN, and DB-SU on the basis of F-measure versus recall. The graph shows that the F-measure versus the recall value shows different behavior for three algorithms. Among three clustering techniques proposed, DB-SU is better, as its F-measure value is increasing with the increase in recall and provides more correct results than K-means or DBSCAN.

5 CONCLUSION AND FUTURE SCOPE

Health care is an important area where a lot of research is occurring due to the requirements of more accurate diagnoses as well as treatments. It is the field in which accuracy is very important. This chapter analyzed traditional DBSCAN used for disease diagnose and found that accuracy in traditional DBSCAN has some disadvantages. So a hybrid algorithm, DB-SU, is proposed to improve the existing DBSCAN results required to overcome the issue. The experimental results show that the proposed algorithm, DB-SU, is providing better and more precise results in terms of accuracy, precision, recall, error rate, and F-measure. The accuracy of the proposed algorithm has an increase of 11.11% as compared to the existing DBSCAN. The precision value and F-measure value of DB-SU has also increased by 35% and 29.4%, respectively, as compared to the existing DBSCAN. In future, the proposed algorithm will be used on many other test data sets that are obtained from medical centers. The other factors and key features of data mining can be analyzed and evaluated based on related methodology.

REFERENCES

JOURNAL ARTICLES

Abdullah, A. S., Rahebi, J., Özok, Y. E., & Aljanabi, M. (2020). A new and effective method for human retina optic disc segmentation with fuzzy clustering method based on active contour model. *Medical & Biological Engineering & Computing*, 58(1), 25–37.

Anand, T., Pal, R., & Dubey, S. K. (2019). Cluster analysis for diabetic retinopathy prediction using data mining techniques. *International Journal of Business Information Systems*, 31(3), 372–390.

Arcadu, F., Benmansour, F., Maunz, A., Willis, J., Haskova, Z., & Prunotto, M. (2019). Deep learning algorithm predicts diabetic retinopathy progression in individual patients. *NPJ Digital Medicine*, 2(1), 1–9.

Arenas-Cavalli, J. T., Ríos, S. A., Pola, M., & Donoso, R. (2015). A web-based platform for automated diabetic retinopathy screening. *Procedia Computer Science*, 60, 557–563.

Bhuvaneswari, C., Aruna, P., & Loganathan, D. (2014). A new fusion model for classification of the lung diseases using genetic algorithm. *Science Direct, Elsevier*, 15(2), 69–77.

Cavan, D., Makaroff, L., da Rocha Fernandes, J., Sylvanowicz, M., Ackland, P., Conlon, J., Chaney, D., Malhi, A., & Barratt, J. (2017). The diabetic retinopathy barometer study: Global perspectives on access to and experiences of diabetic retinopathy screening and treatment. *Diabetes Research and Clinical Practice*, 129, 16–24.

De Carvalho, H. H., Moreno, R. L., Pimenta, T. C., Crepaldi, P. C., & Cintra, E. (2013). A heart disease recognition embedded system with fuzzy cluster algorithm. *Computer Methods and Programs in Biomedicine*, 110(3), 447–454.

Jabbar, A. M., Deekshatulu, B. L., & Priti Chandra (2013). Classification of heart disease using K-nearest neighbor and genetic algorithm. *Science Direct. Elsevier*, 10, 85–94.

Lin, J., Yu, L., Weng, Q., & Zheng, X. (2019). Retinal image quality assessment for diabetic retinopathy screening: A survey. *Multimedia Tools and Applications*, 79(23–24), 16173–16199.

Long, S., Huang, X., Chen, Z., Pardhan, S., & Zheng, D. (2019). Automatic detection of hard exudates in color retinal images using dynamic threshold and SVM classification: Algorithm development and evaluation. *BioMed Research International*, 23, 1–13.

Mwangi, N., Bascaran, C., Ng'ang'a, M., Ramke, J., Kipturgo, M., Gichuhi, S., Kim, M., Macleod, D., Moorman, C., Muraguri, D., Gakuo, E., Muthami, L., & Foster, A. (2020). Feasibility of a cluster randomized controlled trial on the effectiveness of peer: Led health education interventions to increase uptake of retinal examination for diabetic retinopathy in Kirinyaga, Kenya: A pilot trial. *Pilot and Feasibility Studies*, 1–1.

Mwangi, N., Gachago, M., Gichangi, M., Gichuhi, S., Githeko, K., Jalango, A., Karimurio, J., Kibachio, J., Muthami, L., Ngugi, N., Nduri, C., Nyaga, P., Nyamori, J., Zindamoyen, A., & Bascaran, C., Foster, A. (2018). Adapting clinical practice guidelines for diabetic retinopathy in Kenya: Process and outputs. *Implementation Science*, 13(1), 1–9.

Mwangi, N., Ng'ang'a, M., Gakuo, E., Gichuhi, S., Macleod, D., Moorman, C., Muthami, L., Tum, P., Jalango, A., Githeko, K., Gichangi, M., Kibachio, J., Bascaran, C., & Foster, A. (2018). Effectiveness of peer support to increase uptake of retinal examination for diabetic retinopathy: Study protocol for the DURE pragmatic cluster randomized clinical trial in Kirinyaga, Kenya. *BMC Public Health*, 18(1).

Saranya Rubini, S., & Kunthavai, A. (2015). Diabetic retinopathy detection based on Eigen values of the Hessian Matrix. *Science Direct, Elsevier*, 17, 311–318.

Tanabe, H., Saito, H., Kudo, A., Machii, N., Hirai, H., Maimaituxun, G., Tanaka, K., Masuzaki, H., Watanabe, T., Asahi, K., Kazama, J., Shimabukuro, M. (2020). Factors associated with risk of diabetic complications in novel cluster-based diabetes subgroups: A Japanese retrospective cohort study. *Journal of Clinical Medicine*, 9(7), 2083.

Tapas, K., & David, M. M. (2002). An efficient K-means clustering algorithm: Analysis and implementation. Pattern analysis and machine intelligence. *IEEE Transactions on Pattern Analysis and Machine Intelligence*, 24(7).

Conference Proceedings

Ilayaraja, M., & Meyyappan, T. (2013, February). Mining medical data to identify frequent diseases using Apriori algorithm, International Conference on Pattern Recognition, Informatics and Mobile Engineering (PRIME): February 21–22, 2013, Periyar University, Tami nadu, India.

Rajput, G. G., & Patil, P. N. (2014, January). Detection and classification of exudates using k-means clustering in color retinal images. In 2014 Fifth International Conference on Signal and Image Processing (pp. 126–130). IEEE.

Shajahan, B., & Sudha, S. (2014). Hepatic tumor detection in ultrasound images. *IEEE*, 1–5.

Sindhumol, S., Kumar, A., & Balakrishnan, K. (2013). Spectral clustering independent component analysis for tissue classification from brain MRI. *Biomedical Signal Processing and Control*, 8(6), 667–674.

Website

World Health Organization. (2021, January 10). *WHO*. www.who.int/health-topics/diabetes#tab=tab_1

8 Dynamic Simulation Model to Increase the Use of Public Transportation Using Transit-Oriented Development

Rizki Wahyunuari Ningrum, Erma Suryani, & Rully Agus Hendrawan

CONTENTS

1	Introduction	97
2	Literature Review	98
	2.1 System Dynamics Simulation to Increase the Use of Public Transportation and Reduce Traffic Congestion	98
	2.2 Scenario Planning to Increase the Use of Public Transportation and Reduce Traffic Congestion	98
3	Model Development	99
4	Validation	100
5	Scenario Development	101
	5.1 Mass Rapid Transit Scenario	101
	5.2 Light Rail Transit Scenario	101
	5.3 MRT and LRT Scenario	102
6	Conclusion	104
References		105

1 INTRODUCTION

Transportation has become one of the important components for people in carrying out their daily activities. Transportation is a business or activity of moving something from one place to another.[1] Problems such as congestion are common problems that often occur in big cities, one of which is in the city of Surabaya. Congestion is a condition where traffic experiences a decrease in operating speed due to obstacles so that freedom of movement is relatively small.[2] Congestion occurs because the total number of vehicles exceeds the available road capacity, resulting in an accumulation of

DOI: 10.1201/9781003169550-8

vehicles.[3] Every year, there is an increase in the number of vehicles by about 7.03%.[4] Based on data from the Surabaya City Transportation Department, it was found that from 2010 to 2019, the number of vehicle volumes almost always increased, especially the number of private transportations such as cars and motorcycles.

Another factor that causes congestion is the lack of public awareness in using public transportation. This is due to the large amount of public transportation that does not meet the standards and is not feasible so that people choose to use private transportation.[5] The Surabaya City government has tried to overcome congestion by increasing the number of uses of public transportation, one of which is the implementation of the Suroboyo Bus. The Suroboyo Bus has adequate facilities so that it can provide comfort and security so that people are interested in switching to using public transportation.[6] The implementation of the Suroboyo Bus was initially welcomed by the community, but over time, there are still obstacles that cause a lack of effectiveness of the implementation of the Suroboyo Bus. This is because the Suroboyo Bus departure schedule is still not regular, and the service is plagued by long waiting times, inappropriate travel time, and inadequate bus stop facilities.[7]

There are previous studies[8] that adopt a system dynamics model to increase the use of public transportation. System dynamics is a method used to create models of complex real systems.[9] This study uses a system dynamics model to find out what factors can help increase the use of public transportation. The results of the study[8] mention that there is a need to improve facilities, decrease travel time, and add transit-based public transportation. In addition, there are other studies[10] which also mention that the problem of congestion and lack of use of public transportation can be overcome by the development of transit-oriented development (TOD), which supports sustainable transportation modes such as public transportation, walking, cycling, and decreasing travel time in overcoming congestion.

2 LITERATURE REVIEW

2.1 System Dynamics Simulation to Increase the Use of Public Transportation and Reduce Traffic Congestion

The increase in the number of private vehicles causes congestion problems and a decrease in the use of public transportation. The level of traffic congestion is a comparison between the average daily traffic and road capacity with units of passenger car units (PCU) per hour.[11] Meanwhile, the use of public transportation describes the relationship between the volume of vehicles and the number of users of public transportation. If the number of public transportations increases and private vehicle passengers switch to public transportation, the volume of private vehicles will gradually decrease so that it can affect the level of congestion. Vehicle volume decreased slightly due to reduced private transportation.

2.2 Scenario Planning to Increase the Use of Public Transportation and Reduce Traffic Congestion

Scenario development is a method that develops various possibilities using current data to serve as alternative scenarios for various problems.[12] The purpose of the

Dynamic Simulation Model: Public Transporation

simulation is to find out which factors or variables have the greatest influence on the scenario. There are two kinds of scenarios, namely parameter and structure scenarios. Parameter scenarios are scenarios that change the parameters of the variables to get the desired result, while the structural scenario is a scenario that is carried out by adding or subtracting variables that can affect the results of the model to get the desired results.

This study will use a structural scenario so that it can produce the desired output, namely an increase in the number of uses of public transportation and a decrease in congestion levels. The scenario is to add mass-based public transportation such as mass rapid transit (MRT) and light rail transit (LRT). The scenario of adding transit-based public transportation of MRT and LRT is a scenario carried out to increase the number of uses of public transportation by implementing TOD by diverting private transportation users to use MRT and LRT to reduce congestion caused by the number of private vehicles. MRT is a transit-based public transportation that has three main criteria, namely mass (large carrying capacity), rapid (fast travel time with high frequency), and transit (stopping at stopping point stations).[5] Meanwhile, LRT is public transportation that has a large passenger capacity and its own lane. The advantage of LRT is that it does not produce additional pollution while reducing vehicle volume.[13] This scenario is carried out by obtaining the percentage of private vehicle users who are interested in switching to the MRT and using TOD factors such as accessibility and mixed land use which can help increase the growth of public transportation users.

3 MODEL DEVELOPMENT

The system model of a causal loop diagram (CLD) can be seen in Figure 8.1. The causal loop diagram shows that there are two balancing loops and one reinforcing loop. B1 explains that the use of public transportation with the availability of public

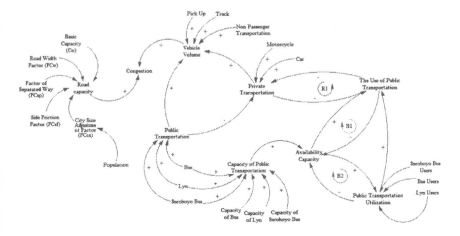

FIGURE 8.1 Causal loop diagram of the Surabaya City Transportation System.

transportation capacity will be balanced: if the use of public transportation increases, the availability of public transportation capacity will decrease, and if the use of public transportation decreases, the availability of public transportation will increase. B2 explains that the availability of public transportation capacity with Public Transportation Utilization will be balanced: if the availability of public transportation capacity increases, the utilization of public transportation will decrease, and if the availability of public transportation capacity decreases, the utilization of public transportation will increase. R1 explains that the number of private transportations will reduce the use of public transportation, and conversely, the use of public transportation will reduce the number of private transportations.

4 VALIDATION

Model validation is the stage to test whether the model has run well and provides the expected output. The technique used to validate the model is the Behavioural Pattern Test. The Behavioural Pattern Test is a validity test in which two kinds of tests are carried out, namely mean comparison and error variance.

Mean comparison:

$$E1 = \frac{|\bar{S} - \bar{A}|}{\bar{A}}$$

E1 = Mean comparison (%)
S = The average value of the simulation results
A = The average value of the actual data
The model is considered valid if the value of E1 ≤ 5%

Error variance:

$$E2 = \frac{|Ss - Sa|}{Sa}$$

E2 = Error variance (%)
Ss = The average value of the simulation results
Sa = The average value of the actual data results
The model is considered valid if the value of E2 ≤ 30%.

The results of the validation on the Lyn user's variable are the mean comparison of 0.00575% and the error variance of 0.00176%, which indicates that the validation carried out is valid. The results of the validation on the bus user's variable are the mean comparison of 0.00012% and the error variance of 0.00124%, which indicates that the validation carried out is valid. The results of the validation on the Suroboyo Bus user's variable are the mean comparison of 0.99% and the error variance of 0.602%, which indicates that the validation carried out is valid. The results of the validation on the population variable are the mean comparison of 0.76% and the error variance of 0.603%, which indicates that the validation has been valid.

The results of the validation on the motorcycle variable are the mean comparison of 0.00069% and the error variance of 0.00396%, which indicates that the validation has been valid. The results of the validation on the private car variable are the mean comparison of 0.0289% and the error variance of 0.0679%, which indicates that the validation has been valid. The results of the validation on the pick-up variable are the mean comparison of 0.00101% and the error variance of 0.00043%, which indicates that the validation has been valid. The results of the validation on the truck variable are the mean comparison of 0.291% and the error variance of 0.379%, which indicates that the validation has been valid. The results of the validation on the Lyn variable are the mean comparison of 0.0199% and the error variance of 0.00098%, which indicates that the validation has been valid. The results of the validation on the bus variable are the mean comparison of 0.0238% and the error variance of 0.019%, which indicates that the validation has been valid. The results of the validation on the Suroboyo Bus variable are the mean comparison of 0% and the error variance of 0%, which indicates that the validation has been valid.

5 SCENARIO DEVELOPMENT

The scenario development is based on the base model that has gone through the verification and validation process. Scenario development is carried out to increase the use of public transportation and reduce congestion. The scenario used in this final project is a structural scenario. The scenario used in this final project is MRT, LRT, and MRT with LRT.

5.1 Mass Rapid Transit Scenario

The MRT scenario is based on the percentage of users who switch to the MRT, which is around 48.5%. These users are 26% motorcycle users and 60.6% car users.[14] TOD will increase the use of public transportation by 0.3% through two factors, namely mixed land use and accessibility.[15] The impact of the TOD will increase the percentage of use of public transportation or passengers who switch to the MRT consistently.

Figure 8.2 shows that from 2020 to 2040, the trend of using public transportation tends to increase up to more than 100.000 passengers per day, while in Figure 8.3, the trend of congestion tends to decrease.

5.2 Light Rail Transit Scenario

The LRT scenario is based on the percentage of users who switch to the LRT, which is around 34.57%. These users are 45.68% motorcycle users and 25.28% car users.[16] TOD will increase the use of public transportation by 0.3% through two factors, namely mixed land use and accessibility.[15] The impact of the TOD will increase the percentage of use of public transportation or passengers who switch to the LRT consistently.

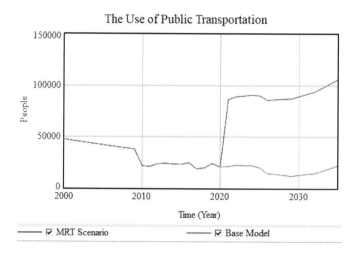

FIGURE 8.2 Graph of MRT scenario on the use of public transportation variable.

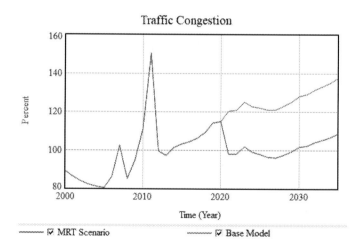

FIGURE 8.3 Graph of MRT scenario on traffic congestion variable.

Figure 8.4 shows that from 2020 to 2040, the trend of using public transportation tends to increase up to 90.000 passengers per day, while in Figure 8.5, the trend of congestion tends to decrease.

5.3 MRT AND LRT SCENARIO

The third scenario is a combination of the two previous scenarios. The combination of the two scenarios will have a greater impact on the use of public transportation and traffic congestion.

Dynamic Simulation Model: Public Transporation

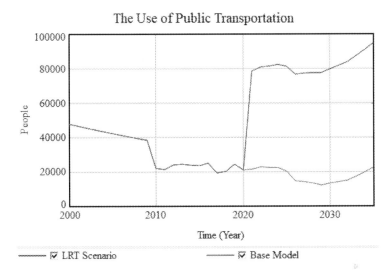

FIGURE 8.4 Graph of LRT scenario on the use of public transportation variable.

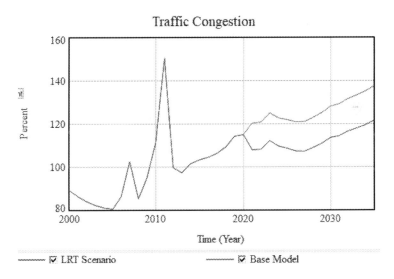

FIGURE 8.5 Graph of LRT scenario on traffic congestion variable.

Figure 8.6 shows that from 2020 to 2040, the trend of using public transportation tends to increase up to 180.000 passengers per day, while in Figure 8.7, the trend of congestion tends to decrease.

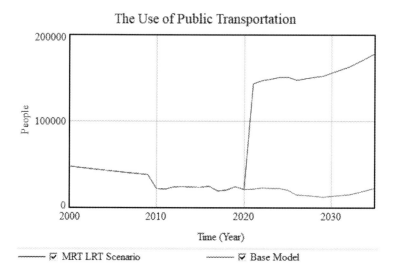

FIGURE 8.6 Graph of MRT and LRT scenario on the use of public transportation variable.

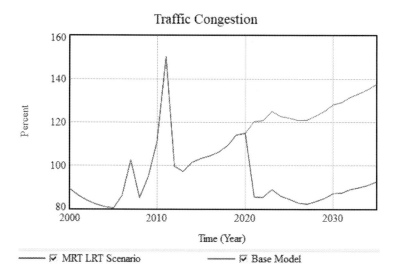

FIGURE 8.7 Graph of MRT and LRT scenario on traffic congestion variable.

6 CONCLUSION

The development of a model to increase the use of public transportation and reduce the level of congestion in the city of Surabaya has been carried out based on current conditions. Based on the model that has been made, it can be said that there is a link between the congestion submodel and the use of public transportation. These factors include private vehicle users who want to switch to using public transportation. To overcome excessive congestion, a scenario is made that can make users of private vehicles such as motorbikes and cars use public transportation such as MRT and LRT. This mode change has a considerable impact on the use of public transportation, causing congestion. The results of the MRT scenario can increase the use of public transportation to exceed 100,000 passengers per day and reduce more than 25% of the congestion rate. The results of the LRT scenario can increase the use of transportation to exceed 90,000 passengers per day and reduce more than 15% of the congestion rate. The results of the MRT and LRT scenarios can increase the use of public transportation to exceed 180,000 passengers per day and reduce more than 40% of the congestion rate.

REFERENCES

1. Kadir A. Transportation: Its role and impact in national economic growth (Tranportasi: Peran dan Dampaknya Dalam Pertumbuhan Ekonomi Nasional). *J Perenc dan Pengemb Wil Wahana Hijau*. 2006;1(3):121–131.
2. SUMADI. Traffic congestion on veterans road (Kemacetan Lalulintas Pada Ruas Jalan Veteran). 2006.
3. Marufuzzaman M, Ekşioğlu SD. Managing congestion in supply chains via dynamic freight routing: An application in the biomass supply chain. *Transp Res Part E Logist Transp Rev*. 2017;99:54–76. doi:10.1016/j.tre.2017.01.005
4. Winaryo MB. The role of the Surabaya city government in regulating Surabaya city public transportation: Urban studies program "Suroboyo Bus" (Peran Pemerintah Kota Surabaya dalam Pengaturan Transportasi Publik Kota Surabaya: Studi Perkotaan Program "Suroboyo Bus"). 2019;8(5):55.
5. de Rozari A, Wibowo YH. Factors causing traffic congestion on main streets in Surabaya (Faktor-faktor Yang Menyebabkan Kemacetan Lalu Lintas di Jalan Utama Kota Surabaya. *J Penelit Adm Publik*. 2015;1(1):1–5. doi:10.1007/s13398-014-0173-7.2
6. Kurniawan AA, Indah Prabawati S. Sos MS. Implementation of the Suroboyo bus at the Surabaya city transportation service (Implementasi suroboyo bus di dinas perhubungan kota surabaya).
7. Risnu DA, Kartika AG, Herijanto W. Analysis of the performance of the Suroboyo bus on the west-east route on the satisfaction of transportation actors (Analisis Kinerja Bus Suroboyo Rute Barat-Timur Terhadap Kepuasan Pelaku Transportasi). *J Tek ITS*. 2019;8(2):20–25.
8. Wicaksono R.DA. Development of system dynamics model to improve the use of public transportation in Surabaya (Pengembangan model sistem dinamik untuk meningkatkan penggunaan transportasi umum di surabaya). 2020. https://repository.its.ac.id/id/eprint/76990
9. Sterman J. System dynamics: Systems thinking and modeling for a complex world. 2002. http://hdl.handle.net/1721.1/102741
10. Ramadhani VS. Gubeng station with transit oriented development concept (Stasiun Gubeng Dengan Konsep Transit Oriented Development). 2017.
11. Direktorat Jendral Bina Marga. Indonesian road capacity manual (Manual Kapasitas Jalan Indonesia). 1997.
12. Schoemaker PJH. Scenario planning: A tool for strategic thinking. *Long Range Plann*. 1995;28(3):117. doi:10.1016/0024-6301(95)91604-0
13. Febrianda M, Herijanto IW. Study on LRT (Light Rail Transit) route planning as a feeder mode for MRT Jakarta (Studi Perencanaan Rute LRT (Light Rail Transit) Sebagai Moda Pengumpan (Feeder) MRT Jakarta). *J Tek Pomits*. 2013;1(1):1–6.

14 Nurfadilah D, Soenarjono B, Alamsjah A, Kautsar M, Perpustakaan TD. Factors that influence customers intention to use MRT. 2020.
15 Zhu Z, Lee M, Pan Y, Yang H, Zhang L. Analyzing the impact of a planned transit-oriented development on mode share and traffic conditions. *Transp Plan Technol.* 2018;41(8):816–829. doi:10.1080/03081060.2018.1526882
16 Sianipar A. Study of community preferences in using the Jabodebek Lrt (Kajian Preferensi Masyarakat Dalam Menggunakan Lrt Jabodebek). *J Penelit Transp Darat.* 2020;21(1):13–20. doi:10.25104/jptd.v21i1.962

9 Text Summarization Using Extractive Techniques

Mukesh Rawat, Mohd Hamzah Siddiqui, Mohd Anas Maan, Shashaank Dhiman, & Mohd Asad

CONTENTS

1	Introduction	107
	1.1 Term Frequency – Inverse Document Frequency (TF-IDF)	108
	1.2 Text Rank Algorithm	108
	1.3 Latent Semantic Analysis (LSA)	109
2	Methodology	109
	2.1 Data Collection	109
	2.2 Data Preprocessing	109
	2.2.1 Tokenization	109
	2.2.2 Normalization	110
	2.3 Term Frequency – Inverse Document Frequency (TF-IDF)	110
	2.4 Text Rank Algorithm	112
	2.5 Latent Semantic Analysis (LSA)	112
	2.6 ROUGE-N Metric	115
3	Result Analysis	115
	3.1 Preprocessing	115
	3.2 Term Frequency – Inverse Document Frequency (TF-IDF)	115
	3.3 Text Rank Algorithm	117
	3.4 Latent Semantic Analysis (LSA)	117
4	Conclusion	118
Note		119
References		119

1 INTRODUCTION

In this new age where a lot of information [1] is developed each moment, there is a need to productively get to the main substance. The lives of present individuals are relentless, and they need everything to be done in a short of time. No one has the opportunity to peruse each line of any text, whether it is news, authoritative reports, messages, books, and so on We need an ad-libbed system (i.e. improvised mechanism) to separate the data quickly and productively in light of the fact that human beings can't do

DOI: 10.1201/9781003169550-9

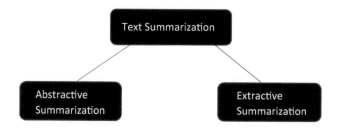

FIGURE 9.1 Types of text summarization.

this without anyone else to extricate the synopsis of enormous reports. There is an abundant measure of information accessible on the web, so there is an issue of finding the proper documents from the corpus of information and recovering applicable data.

To redress this issue, a book synopsis is required. Text summarization alludes to contracting the given content into a more modest rendition while monitoring its data information and meaning. It makes a short and precise rundown of a long report. This strategy encourages us to discover proper data and to devour the applicable data rapidly.

We have developed a substance diagram structure using both extractive and abstractive methodologies. We will embrace three computations in each methodology, and a short time later, we will survey their outcomes. For this undertaking, we have focused in on extractive philosophy (term frequency-inverse document frequency, text rank, latent semantic analysis).

Abstractive summarization: The abstractive approach involves summarization based on deep learning. It uses new phrases and terms, different from the actual document, keeping the points the same, just like how we actually summarize, so it is much harder than the extractive approach.

Extractive summarization [2]: The extractive approach involves picking up the most important phrases and lines from the documents. It then combines all the important lines to create the summary. In this case, every line and word of the summary actually belongs to the original document which is summarized.

1.1 Term Frequency – Inverse Document Frequency (TF-IDF)

Term frequency (inverse record frequency) is a programmed synopsis dependent on weighting. It considers sentences to be arranged successions and words as requested subsets of sentences. It has four stages: weighting of words, weighting of sentences, picking all sentences over a specific weight limit, and amassing the chosen sentences as they show up in the first article.

1.2 Text Rank Algorithm

The text rank algorithm is a solo chart-based positioning calculation that utilizes the intuition behind the page rank calculation. The diagram-based positioning

Text Summarization: Extractive Techniques 109

calculations choose the noteworthiness of a vertex by considering data about the entire chart instead of the vertex of distinct data. It makes a diagram utilizing some arrangement of text units as vertices for key phrase extraction.

1.3 Latent Semantic Analysis (LSA)

The latent semantic analysis technique picks one sentence for every point. It relies upon the length of overviews in relationship to sentences while holding the quantity of topics. One of the updates was to pick the length of the once-over dependent on the greatness of the point, which urges us to have a variable number of sentences. It was comprehended that the sentences which are basic for huge subjects should be accessible in the summary. In this manner, to discover these sentences, weight was connected with each sentence.

2 METHODOLOGY

2.1 Data Collection

The synopsis framework or the summarization system is language-subordinate; it takes a shot at English archives. We have taken information from two sources:

1. Web links of online journals
2. Downloaded reports saved money on the client's machine as .txt records.

The client can place any English archive in the framework as info and get the summed-up yield utilizing any of the proposed procedures.

2.2 Data Preprocessing

There are various preprocessing systems [3] that can be used on chronicles for summation like disposing of stop words, normalizing terms, replacing counterparts, etc. We are using NLTK[1] close by standard Python libraries. The Natural Language Toolkit is remarkable contrasted with other known and most-used NLP libraries in the Python climate, supportive for endeavors from tokenization [4] to stemming to linguistic element marking, etc. Preprocessing of data involves following steps.

2.2.1 Tokenization

A greater size of text can be tokenized into sentences, sentences can be tokenized into words, etc. Tokenization is also called text division or lexical assessment. The division is, moreover, every so often used to insinuate the breakdown of a tremendous chunk of text into pieces greater than words (for instance, sections or sentences). However, tokenization puts something aside for the breakdown strategy, which closes absolutely in words. We are tokenizing our example text into a once-over of words utilizing NTLK's statement tokenize () work. words = nltk.wordtokenize (sample).

Given a document D, we tokenize it into sentences as $\langle s_1, s_2, s_3, s_4...s_n \rangle$.

2.2.2 Normalization

Normalization is changing all content over to a similar case (upper or lower), eliminating accentuation, changing numbers over to their word equivalent, and so forth. It puts all words on equivalent balance and permits preparing to continue consistently.

1. *Channel out punctuation.* We can eliminate all tokens that we are not keen on, for example, all independent accentuation. This should be possible by repeating overall tokens and just keeping those tokens that are, on the whole, alphabetic. Python has the work isalpha() that might be utilized.
2. *Channel out stop words (and pipeline).* Stop words are those words that don't have noteworthiness in forming the record. They are the most notable words, for instance, "the", "a" and "is". For some applications like documentation classification cation, it's going to be to get rid of stop word. NLTK gives a posting of typically supported stop words for a spread of tongues, like English.

2.3 Term Frequency – Inverse Document Frequency (TF-IDF)

Term frequency (inverse document frequency) [5] is a programmed synopsis dependent on weighting. It considers sentences to be arranged groupings and words as requested subsets of sentences. It has four stages: weighting of words, weighting of sentences, picking all sentences over a specific weight limit, and amassing the chose sentences as they show up in the first article.

This technique for weighting depends on frequencies. Each word, called a term, is designated weight utilizing the tf-idf (term frequency {inverted document frequency) approach.

Weight of a term = term frequency × inverse of document frequency

Term frequency is the check of a word that happens inside a document. Inverse document frequency is 1/number of documents the words show up in.

We can take archives from the web: to do so, we can use web-scratching libraries like Delightful Soup, or we can fundamentally download some substance from the web and save it as .txt rec-ord.

By and by, for finding the recurrence of occasions of each word, a coordinated article text variable is used. This variable is used to pay special mind to the recurrence of regularity since it doesn't contain punctuation, digits, or other exceptional characters. We keep on envisioning whether the words exist in the word recurrence word reference – for instance, word frequencies – or not. If the word is encountered for the first time, it is added to the dictionary as a key and its value is set to 1. Something different: if the word earlier exists inside the jargon, its value is basically increased by 1. Finally, to find the weighted recurrence, we can basically isolate the quantity of occasions of the large number of words by the recurrence of the most common word. We have at present decided the weighted frequencies for all the words. The present time is an ideal occasion to learn the scores for each sentence by adding weighted

Text Summarization: Extractive Techniques

frequencies of the words that occur in this unequivocal sentence. As of now, we have the sentence scores lex-image that contains sentences with their looking at score. To summarize the article, we will take high-N sentences with the most awesome scores.

Algorithm to Calculate idf

```
Input: term: Term in the Document,
       allDocs: List of all documents
Return: Inverse Document Frequency (idf) for term
       = Logarithm ((Total Number of Documents) /
         (Number of documents containing the term))

# Iterate through all the documents
Putting a check if a term appears in a document.
If term is present in the document, then
increment "check" variable
```

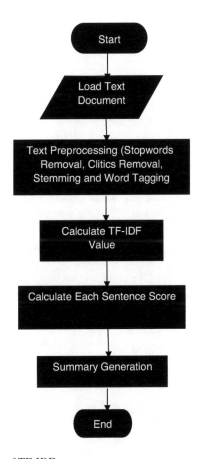

FIGURE 9.2 Flowchart of TF-IDF.

```
# Calculate Total number of documents

# Calculating the IDF
idf_val = log(float(total_num docs)/num_docs_with_
   given_term)
```

The flowchart in Figure 9.2 represents the working of TF-IDF.

2.4 TEXT RANK ALGORITHM

The text rank algorithm [6] is a solo diagram-based positioning calculation that utilizes the sense behind the page rank calculation. The graph-based situating computations pick the immensity of a vertex by considering information about the whole chart rather than the vertex of particular information. It makes a diagram using some course of action of text units as vertices for key-phrase extraction. Here, we use sentences as the vertices and words as the edges. Edges depend on some extent of semantic or lexical likeness between the substance unit vertices. It changes over the sentences of the initial archive into vectors and determines their similarity [7] for a once-over. We use the cosine closeness where we convert sentences to vectors to enlist comparability. This estimation is sans language. Steps are as follows:

1. Produce a term-archive network (TD grid) of the data: Conversion of text reports into a structure of token checks. CountVectorizer() is being used. A 2D organization containing the record text framework is created therefore. A graph to proceed for the textrank estimation: nxgraph is a chart made using the association library. Each sentence having fundamental words is addressed by a center. The amount of words that are standard in both of the sentences(nodes) shows up as weight edges. nx.draw() strategy is being used to draw the graph made. Finding critical sentences and creating plot: sentencearray is the organized (dropping solicitation w.r.t. score regard) 2D show of ranks[sentence] and sentence. For example, if there are two sentences: S1 (with a score of S1 = s1) and S2 with score s2, with s2 > s1, then sentencearray is [[s2, S2], [s1, S1]]
2. Computation of limit: We take the mean estimation of standardized scores.
3. Separate out the sentences that fulfill the rules of having a score over the edge weight.

2.5 LATENT SEMANTIC ANALYSIS (LSA)

Latent semantic analysis [8] is a solo technique dependent on noticed words to extricate the portrayal of text semantics. From the outset, this procedure shapes a term sentence lattice (*n by m network*), each line addresses a word from the data (*n* words), and each section addresses a sentence (*m* sentence). In the organization, each part is put by *ij* structure, for instance, weight of word i in sentence *j*. TF-IDF methodology calculates the greatness of words, and if a word is missing in the sentence, the weight of that word in the sentence is zero. Another methodology called specific

Text Summarization: Extractive Techniques

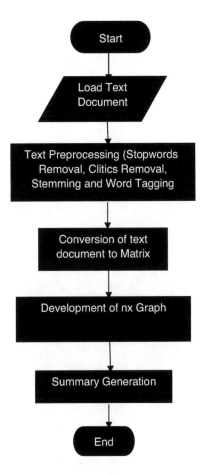

FIGURE 9.3 Flowchart of text rank algorithm.

worth crumbling (SVD) is applied on the lattice, which changes system An into three organizations: A = U V T

Lattice U ($n \times m$) = term subject framework with loads of words
Framework (slanting lattice $m \times m$) = each line I relates to weight of point I
Network V = subject sentence framework
Grid D = V T-characterize how much a sentence speaks to a point. Thus, d_{ij} shows weight of point i in sentence j

The lethargic semantic examination procedure picks one sentence for each subject. It relies upon the length of overviews in relationship to sentences while holding the quantity of subjects. One of the redesigns was to pick the length of the diagram dependent on the greatness of the point, which makes us have a variable number of sentences. It was valued that the sentences which are basic for huge topics should be accessible in the summary. Accordingly, to discover these sentences, weight was connected with each sentence.

Algorithm for Latent Semantic Analysis

```
from sklearn.feature_extraction.text import
            TfidfVectorizer
 from sklearn.decomposition import TruncatedSVD
 from sklearn.pipeline import Pipelinedocuments =
["Document_A.txt", "Document_B.txt", "Document_
            C.txt"]
 #Raw documents to tf-idf matrix (or normal count
            could be done too)
vectorizer = TfidfVectorizer(stop_words='english',
            use_idf=True,
            smooth_idf=True)
      #SVD for dimensionality reduction
svd_model = TruncatedSVD(n_components=100,
            // num dimensions
                algorithm='randomized',
                    n_iter=10)
#Pipe tf-idf and SVD, apply on our input documents
 svd_transformer = Pipeline([('tfidf', vectorizer),
('svd', svd_model)])svd_matrix = svd_transformer.
            fit_transform(documents)
```

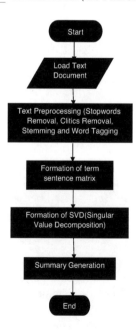

FIGURE 9.4 Flowchart of latent semantic analysis.

Text Summarization: Extractive Techniques 115

2.6 ROUGE-N Metric

ROUGE-N, or Recall-Oriented Understudy for Gisting Evaluation [9], implies a word N-gram measure between the gold diagram and the model. It is the extent of the check of N-gram states that appear in both gold summation and the model, to the check of all N-gram phrases present in the gold overview. It can in like manner be portrayed as the re-see regard that evaluates the quantity of N-grams from the gold overviews found in the model abstracts.

It is measured on the comparison between the machine-generated output and the reference output based on N-grams.

For overview appraisal, ROUGE-1 and ROUGE-2 estimations are used. As we increase the N, we extended the length of the N-gram word state expected to organize in both gold and model outline.

Model – We have two semantically like expressions: "apple bananas" and \ bananas apples". At the point when we use ROUGE-1, we just observe the unigrams, which are the equivalent for the two expressions. At the point when we use ROUGE-2, we utilize a two-word state; then \apples bananas" becomes unique in relation to "bananas apples", prompting a lower assessment score.

Gold Summary = A great eating routine must have apples and bananas.
Model = Apples and bananas are must for a decent eating regimen
ROUGE-1 = 7/8
ROUGE-2 = 4/7

The proportions reveal to us the measure of data separated by our calculation, which is actually the meaning of review. Subsequently, ROUGE is review based.

3 RESULT ANALYSIS

The consequences of the different advances appear in the figures 9.6 to 9.8.

3.1 Preprocessing

Figure 9.5 shows the aftereffect of the preparing of information. The words have been tokenized and standardized.

3.2 Term Frequency – Inverse Document Frequency (TF-IDF)

To calculate TF-IDF, we use:

```
tf_idf = {}
for i in range(N):
  tokens = processed_text[i]
  counter = Counter(tokens + processed_title[i])
  for token in np.unique(tokens):
    tf = counter[token]/words_count
```

```
df = doc_freq(token)
idf = np.log(N/(df+1))
tf_idf[doc, token] = tf*idf
```

Figure 9.6 shows the aftereffect of TF-IDF. It is the summary generated by TF-IDF using weights.

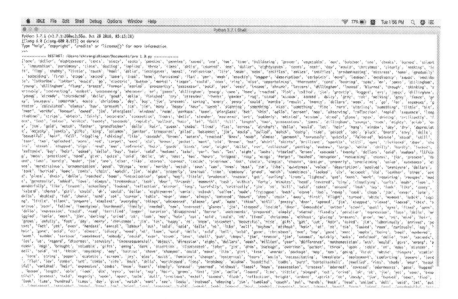

FIGURE 9.5 Result after preprocessing of data.

"A phone rings and text appears on the screen: "Call trans opt: received.
Trinity discusses some unknown person.
Cypher taunts Trinity, suggesting she enjoys watching him.
She ends the call.
It's Trinity.
She brings her hands up from the laptop she's working on at their command.
Agent Smith and the presiding police lieutenant argue.
Agent Smith replies: "No, Lieutenant.
Trinity makes a daring leap across an alley and through a small window.
The phone begins to ring.
The three Agents reunite at the front of the truck.
There is no body in the wreckage.
"She got out," one says.
The other says, "The informant is real."
"We have the name of their next target," says the other, "His name is Neo."
"The Matrix has you."
Trinity approaches him and introduces herself.
Back at his bleak cubicle Neo receives a delivery as "Thomas Anderson."
Upon opening the package he finds a cellphone which immediately rings.

FIGURE 9.6 Summary generated by TF-IDF.

Text Summarization: Extractive Techniques 117

3.3 TEXT RANK ALGORITHM

Figure 9.7 shows the aftereffect of text rank. It is the synopsis created by text rank utilizing vertex portrayal.

3.4 LATENT SEMANTIC ANALYSIS (LSA)

Figure 9.8 shows the aftereffect of LSA. It is the summary generated by LSA.

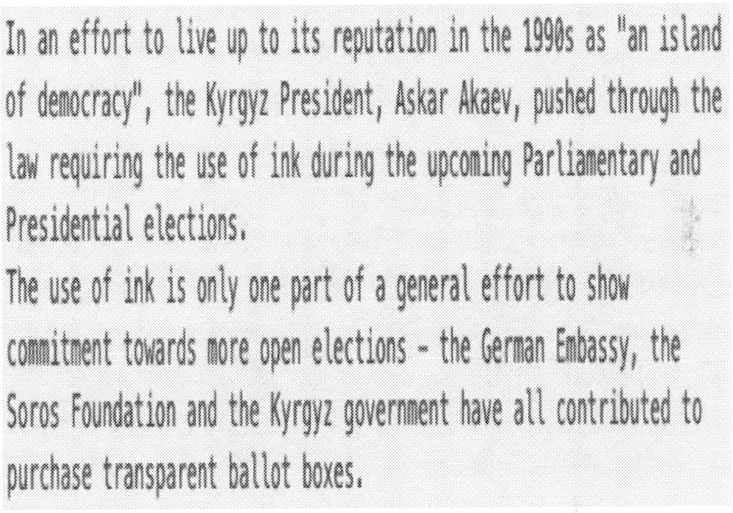

> In an effort to live up to its reputation in the 1990s as "an island of democracy", the Kyrgyz President, Askar Akaev, pushed through the law requiring the use of ink during the upcoming Parliamentary and Presidential elections.
> The use of ink is only one part of a general effort to show commitment towards more open elections – the German Embassy, the Soros Foundation and the Kyrgyz government have all contributed to purchase transparent ballot boxes.

FIGURE 9.7 Summary generated by text rank.

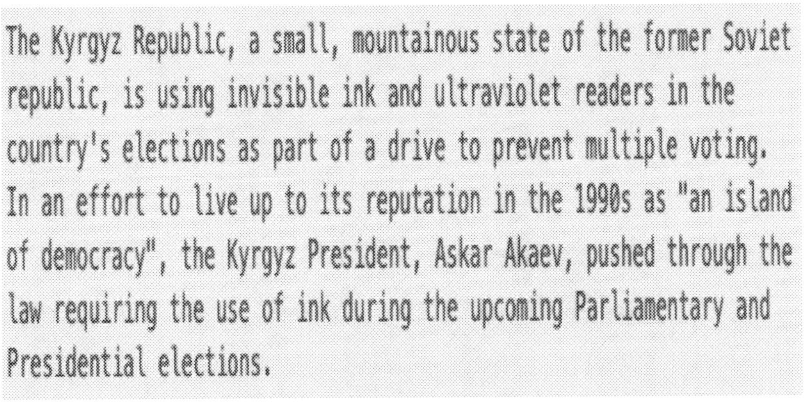

> The Kyrgyz Republic, a small, mountainous state of the former Soviet republic, is using invisible ink and ultraviolet readers in the country's elections as part of a drive to prevent multiple voting. In an effort to live up to its reputation in the 1990s as "an island of democracy", the Kyrgyz President, Askar Akaev, pushed through the law requiring the use of ink during the upcoming Parliamentary and Presidential elections.

FIGURE 9.8 Summary generated by LSA.

4 CONCLUSION

ROUGE-1 is the proportion of covers of unigrams between the framework and reference synopsis. The ROUGE-1 score is determined as the proportion of covers of unigrams among framework and reference rundown, as clarified in the recently given model.

In Table 9.1, we have addressed the ROUGE-1 score for three unmistakable procedures, specifically, TF-IDF approach, text rank approach, and LSA approach. Assessment of ROUGE-1 estimation lies someplace in the scope of 0 and 1. Here, 0 to 1 is the extent of accuracy where 0 is least careful and 1 is the most exact. From Table 9.1 we can draw that, as the ROUGE-1 score [10] of LSA is nearest to 1, likewise, the summary delivered using LSA approach is more exact than the blueprint made using the text rank approach, which is better than the once-over created using the TF-IDF approach.

Consequently, the request for positioning can be closed as:

1. Latent semantic analysis (LSA)
2. Text rank
3. Term frequency-inverse document frequency (TF-IDF)

Figure 9.9 is a reference diagram that speaks to the absolute information esteems created by the ROUGE-N metric.

TABLE 9.1
Accuracy comparison table using ROUGE-N metric

Model	ROUGE-1 Score
TF-IDF	0.197
Text Rank	0.211
LSA	0.230

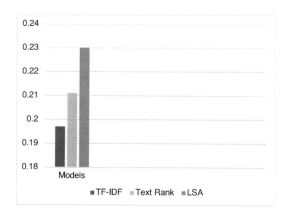

FIGURE 9.9 Bar graph of ROUGE-N metric.

NOTE

1 NLTK is a suite of libraries and programs for symbolic and statistical natural language processing for English written in the Python programming language.

REFERENCES

[1] Jandsalar, H., P. Kumar, and M. Rawat, "An Enhanced Boolean Retrieval Model for Efficient Searching", *Scientific Journal of India*, 2017, 2(1), 21–23.
[2] Deepali, K. Gaikwad, and C. Namrata Mahender, "A Review Paper on Text Summarization", Department of C.S. & I.T., Dr. B. A. M. U., Aurangabad, Maharashtra, India, March 2016.
[3] Savelieva, Alexandra, Bryan Au-Yeung, and Vasanth Ramani, "Abstractive Summarization of Spoken and Written Instructions with BERT", In KDD Converse 2020, August 2020.
[4] Agarwal, N., M. Rawat, and V. Maheshwari, "Comparative Analysis of Jaccard Coefficient and Cosine Similarity for Web Document Similarity Measure", *International Journal for Advance Research in Engineering and Technology*, 2014, 2(10), 18–21.
[5] Bhatia, Neelima, and Arunima Jaiswal, "Automatic Text Summarization and It's Methods: A Review", 6th International Conference, January 2016.
[6] Aggarwal, G., and M. Rawat, "Ranking of Web Documents for Domain Specific Database", *International Journal of Computer Applications*, 2016, 135(6), 16–18.
[7] Sharma, A., and M. Rawat, "Comparison and Analysis of Two Approaches to Find Novel Documents Out of Several Documents", *International Journal of Computer Applications*, 2016, 975, 8887.
[8] Allahyari, Mehdi, "Text Summarization Techniques: A Brief Survey Computer Science Department", University of Georgia, July 2017.
[9] Moratanch, N., and Chitrakala Gopalan, "A Survey on Extractive Text Summarization", Anna University, January 2017.
[10] Jain, S., and M. Rawat, "Efficiency Measures for Ranked Pages by Markov Chain Principle", *International Journal of Information Technology*, 2020, 1–8.

10 An Efficient Deep Neural Network with Adaptive Galactic Swarm Optimization for Complex Image Text Extraction

Digvijay Pandey & Binay Kumar Pandey

CONTENTS

1. Introduction 121
2. Review of Literature 123
3. Problem Statement and Motivation 127
4. Proposed Methodology 128
   - 4.1 Evaluation Metrics 128
   - 4.2 Comparative Analysis 132
5. Conclusion 133
References 133

1 INTRODUCTION

Complex image documents are recognized as online and offline accessible and important precious media which contains vital useful information. These images consist of pixels; after that, the important information from complex images is fetched as per the requirement of computer vision [1]. Textual messages from complex videos and complex images contain accurate meaningful messages, same as the textual information available in images that are used by many complex electronics learning and indulgent implementations like language translator, digitization of book, recovery of video or image, etc. [2,3]. In the latest trend, the use of a pre-program text detection process followed by text recognition has experienced enormous demand. In the past, a lot of research processes are available on removal of text from complex scene image, while this text acquisition process is recognized as a one of the important portions of optical character recognition (OCR) [1,4,5]. OCR commercial processes are used after performing text detection and further binarization to perform the recognition of text from images [6].

DOI: 10.1201/9781003169550-10

Scene images [7] offer suitable and accurate information for blind direction-finding and scene perceptive and recovery approaches, respectively [8,9]. A diversity of fonts and other properties are often built into this complex image [10]. In a complex degraded image, the main focus may include distinctly designed characters, information exposed in digital signposts displayed on a monitor. This is a very typical task for the traditional OCR to identify textual information with different appearances. In this degraded image, the text is scattered around on a frequent basis and the preceding information concerning their position is not presented. The input documents from camera recognize the line spacing, numbers, and characters, but the complex image text does not comprise any formatting rules, so it is not possible to directly introduce the segmentation approach to complex images [11].

In the complex image, the process of text retrieval from the image is precisely predictable by the features of the image. The vision to accomplish high exactness is vastly improved by the heterogeneous features of the image [12]. The process of text detection from complex degraded images contains two main steps: text detection and recognition [13,14]. The key idea of these two processes in complex degraded images is to identify and position the textual information present in complex video or images. This text extraction procedure is executed in many applications like fetching of images and videos from the DBMS (database management system), data processing multimedia retrieval, understanding the natural complex degraded image connotation, and monitoring of traffic [15].

The text extraction procedure is assessed into three main techniques based on connected component, texture detection, and edge detection [16]. The connected component–based method does the quantization of color in an image and region expansion to make clusters of nearest pixels having the same colors in the connected components. An entire structure of every character isn't held in reserve by the connected components because of color bleeding and very low contrast available in each row of textual data. As result of this method, which is based on connected, is set up to be not suitable for image frames obtained from video. Therefore, approaches based on edge detection are generally replaced by a texture-based (TB) method. In texture-based methods, unique text plays a key role to identify the text region. These methods then apply wavelet decomposition (WD); fast Fourier transform (FFT), Gabor transform (GT), and discrete cosine transform (DCT) to extract the features. Normally, these methods apply classifiers for fast neural networks (NN) and SVM (support vector machine) [17,18]. In this [A] presents a new scheme for compression of images in both the time and frequency domains, and it includes wavelet transform to identify a subband of an image and then further decompose into certain levels. For encoding, the wavelet, along with a noise-shaping bit allocation method, was used that may further consider details of less visible images available at high resolution.

The vital procedure performed by the complex degraded image text identification is extraction region of candidate character available in the image. The most important purpose of this region classification method is to obtain the nontext parts from the extracted CCRs [19]. Both the recognition of text and removal of text from the image are considered to be more basic and efficient for various text identification–based applications [20]. The two processes discussed are difficult due to the existence of some factors:

- **Diversity of text in scene:** In contrast to text in images, which are generally with font, single colour, regular size, and regular arrangement, texts in images may allow fully different fonts, colours, scales, and orientations, even in the same image.
- **Complexity of background:** The nontext region in complex degraded images and videos can be very multifarious because it contains elements like signs, bricks, and grasses that are practically difficult to distinguish from true text and thus generate confusion and inaccurate results.
- **Interference factors (nonuniform illumination):** Various interference are like noise, blur, distortion, and low resolution may give rise to failures in image text detection and recognition.

Although, some recently developed methods achieve a high accuracy for text extraction, attaining satisfactory results, such degraded images are still considered a challenging issue [21].

The main work of extraction of text from image is mostly used in various applications in real time. The texts which are removed from images are usually utilized for obtaining the details for guiding tourists, detecting car parking place, visually impaired people, etc. An efficient deep neural network (DNN) is introduced during this method with the objective of text extraction. Before that, the image frame must be analysed to know whether the available image holds any relevant details or not. In order to do it, an approach based on Weighted Naïve Bayes Classifier (WNBC) is introduced before the use of character recognition based on deep learning. WNBC basically distinguishes the textual image from the nontextual image. The error occurs due to reclassification during the text classification process is further streamlined via the use of an optimization method. Lastly, the image that contains the text is input to the DNN-AGSO algorithm for text extraction. In addition to this, during the text extraction process, the load parameter available in the DNN also reduces the rate of accuracy in the DNN. To keep away from such situations, a hybrid approach along with adaptive optimization and DNN is used, which further obtains an optimal weight parameter for the DNN.

2 REVIEW OF LITERATURE

Some of the related works that carry out the extraction of text from natural scene images are discussed in what follows.

A new Mask R-CNN (region-based convolutional neural network) based text recognition technique was developed to identify curved and multi-oriented text in a unified form [22] using complicated degraded images. A new approach for text recognition [23] and text detection was a CC-based approach that utilized the maximally stable extremal regions. The multiple blur produced by motion and defocus makes the text detection process a challenging one. In this [24], a method for text identification in a distorted or non-distorted images is discussed, and the contrast variants experienced in nearby pixels were identified in this method to evaluate the blur degree; moreover, the low-pass filter was used for de-blurring. Mostly, this approach gives pixels under consideration for the purpose of de-blurring images. The process of detecting the scene text from videos attains high value in various content removal–based video applications like video recovery and investigation. This [25] gives text tracking and recognition approaches for frames of videos. The public scene text video was included in this method, which outperforms the other existing methods.

An image compression method with the feature of embedded code is described in [26]. The embedded code denotes a chain of binary decisions that helps to discriminate a meaningful image from the "null" image. Additionally to giving a full bit stream of embedded code, Embedded Zerotree of Wavelet (EZW) results sufficient for comparison with known compression algorithms based on standard test images.

In [27], a wavelet filter–based filtering technique is presented that contains nice resolution in the region-of-interest matching and then reduces resolution in the periphery. This type of method denotes the multiple region of interests (ROIs) that further improved earlier work on MIP-mapping to the domain of the wavelet. Degradation in an image is obtained through scaling of the wavelet coefficient and then Voronoi portioning in the image.

This paper [28] includes region-of-interest–based coding functionality along with the algorithms of set partitioning in hierarchical trees (SPIHT) that is further used for wavelet-based image coding [29]. By giving more focus to the transform coefficients related to the ROI, the ROI is coded with greater constancy then the rest of the image in initial stages of progressive coding. In this research, the main focus is to retrieve the main coefficients in the wavelet-transform domain for further input to the decoder to reconstruct the desired region. This new method gives improved outcomes compared with earlier methods.

In [30], the author presents an improved algorithm to capture textual data in still image or interactive video frames with more complex backgrounds along with image size reduction. This algorithm basically retrieves the line of textual data on the basis of study done on edges, baseline of location, and heuristic constraints. After that, the support vector machine method is applied to obtain a line of textual data from the available candidates in an edge-based distance map of feature space. The experiments performed on plenty of still images and interactive video frames collected from various sources showed the good result of this algorithm as compared to other algorithms.

In [31], a work on image improvement using α-trimmed mean filters is proposed. Image feature improvement is considered as one of the difficult preprocessing tasks in various applications in the processing of an image. At the present time, various methods such as median filter, α-trimmed mean filter, etc. are used in processing of images. But the α-trimmed mean filter is considered as the modification of various filters such as median or mean filters. The proposed algorithm by this author has shown remarkable performance in reducing noise.

In [32], binarization of adaptive documents is proposed, where a single document is assumed as a group of components such as text, background, and picture. In this work, a new way that firstly does quick classification of the local contents of a page to background, pictures, and text is completed. After that threshold the value of each pixel is decided based on two different methods: the soft decision method and the text binarization method. A soft decision method is used for background and foreground, and a unique text binarization method is used for differentiating between text and line drawing areas. The soft decision method does filtering of noise and tracking of signal, while the text binarization method is used to identify text from a degraded background image due to uneven illumination or noise. In the final step, all output of these algorithms is summed together.

This paper [33] suggested the challenging problem of an automatic method for retrieval of information from colored images. However, in order to create databases of bibliographic data, editing is usually performed to provide information related to the titles, etc. For the process automation of indexing, text element identification is considered as an important process. In this paper, the author proposes two methods for automatic text extraction from colored documents and finally a combined outcome of both methods to distinguish between textual and non-textual elements.

In this [34], the author discussed the use of information related to edges for identifying textual blocks from greyscale images. Its main objective is detecting text on noisy images and differentiating them from graphical images. In this, an algorithm is made to obtain the features from various objects and then make a class of those feature points for identifying textual regions. By using methods such as line of approximation and categorization of layout, directionally placed text block can easily obtained. In the final step, merging of feature-based connected components is performed to gather homogeneous textual regions within the scope of its bounding rectangles. The method proposed here gives promising results that show effectiveness.

In [35], the author discusses a binarization technique for color images, and it is found that the traditional method that is based on thresholding does not give better results for an image mixed with foreground colors and background colors. Firstly, on the basis of luminance distribution, features of the image under consideration are obtained. Afterwards, binarization is done with the help of a method based on a decision tree that select different features of color image to binarize images. During the process of feature extraction of a color image, if it is found that colors in the image are intense within a defined color range, then an effective saturation is performed on the image. And if the foreground colors are more dominant, luminance is considered as one important parameter. Finally, if the colors of the background of the image are intense within a defined range, luminance is also applied, and if the number of pixels with low luminance property is less than 60, then saturation may applied; otherwise, both luminance and saturation are applied. The experiments explained in this work include a total of 519 color images, and most of them are invoice and name-card document images. In this work, it is found that the binarization method proposed gives better results as compared to others in terms of shape and connected components.

In [36], detection of text in images frame or videos is supposed as an essential step for retrieval of any multimedia information. In this paper, the author proposed an improved algorithm that can identify, localize, and extract horizontally aligned text in images from degraded backgrounds. The proposed approach depends upon techniques for the reduction of a color, an edge-detection–based method, and the text regions localization using projection profile analyzes geometrical properties of color images. The algorithm outputs consist of various text boxes with a very simple background that are prepared to input into an OCR engine for consequent recognition of character. The performance of the approach is shown by giving promising experimental results for a set of images, frames, and videos.

In [37], the author explained a method to detect texts in images using a texture-based method. The textural properties of texts are analysed using support vector machine. In this method, no different method is used for retrieval of features of textual data; instead, the value intensities of the raw pixels that make a textual pattern are given to the SVM-based classifier. In another method, a continuously adaptive mean shift algorithm is applied for analysis of textual data to identify textual regions. The amalgamation of SVMs and Continously Adaptive Mean Shift Algorithms improved text detection results.

In [38], the author gives details about textual data present in images and video frames for annotation, indexing, and structuring of images. Retrieval of such information from images contains processes like identification, localization, tracking, extraction, enhancement, and recognition of the text data from an image. On the other hand, variations of textual data because of various parameters may generate problems in an automatic method of text extraction.

In [39], the author explained that textual data available in both still images and interactive video frames carries important details for developing better knowledge about optical content and its outcome. In this paper, the author explains uses of multiple wavelet features and gives a new coarse-grain-to-fine-grain–based function that is able to recognize textual data in the highly distorted background of an image. In the first step, the coarse detection–based algorithm is applied. After that, the energy feature of the wavelet is computed to locate all possible pixels that represent text. Thereafter, a method based on region growing is applied to connect all these pixels into regions further filtered into lines of text from available structural information. Secondly, in the fine-grain detection method, four different methods for feature extraction of textured images are issued to represent the texture pattern of the text line to select the most effective features. Then finally, an SVM classifier is used to classify text from the available image database on the selected features. Results of this work were found to be robust for identification of textual data in various conditions.

In [40], the author gives a two-phase noise-removing scheme based on a two-phase noise-removing technique from images like salt and pepper. In the first phase, a pixel is identified that is most likely affected by noise using an adaptive median filter. In the second phase, an image is again restructured using a specific regularization function applied to the selected noisy images. In terms of parameters such as edge perpetuation and noise suppression, their regained images give a significant improvement as compared to nonlinear filters.

In [41], the author proposed that the local wiener filtering along with wavelet domain is an effective image denoising method for less degraded images. In this paper, the author proposes a doubly local wiener filtering algorithm–based method, which uses elliptic directional windows for various sub bands to perform calculation on the variances of signal for noisy wavelet coefficients, and the two other procedures of local wiener filtering are performed on the noisy image. The results obtained after the experiment show that the algorithm proposed in this work may improve the denoising performance.

In [42], the author presents a novel adaptive approach for the binarization as well as improvement of degraded images. The method described does not show any requirement parameter that is changed by the user and may easily handle distortion in an image, which normally happens because of shadows, uneven illumination, less contrast, extremely signal-dependent noise, smear, and strain.

In [43], the author describes a method that is based on sparse representation of signals called morphological component analysis (MCA). Morphological component analysis relies on the assumption that for every signal, atomic behavior must be separated, and there exists a dictionary that makes its construction with the help of sparse representation. After this, the desired separation can be obtained using a pursuit algorithm for the sparse representation. The paper also includes several application results obtained on image content, some theoretical results that give explanation for the separation process.

In [44], the author presents a novel method that is focused on the addition of the basis pursuit denoising (BPDN) algorithm along with a total-variation (TV) regularization scheme used for the separating of texture features and cartoon parts from the image. The author suggests using two dictionaries for the representation of textures and natural scene parts, respectively. Both dictionaries prompt sparse representations of the image over single image content. The main use of the basis pursuit denoising gives a method for the desired separation as well as noise removal. The separation process is directed using the TV regularization scheme; also, it is used for removing ringing artifacts. A highly improved numerical scheme describes a method to provide a solution for a combined optimization problem and several experimental results that validated performance the proposed algorithm.

In [45], the author proposed modeling of textured images by minimization of function and partial differential equations. In this work, an image is decomposed into a summation of two functions represented by u+v where u represents a bounded variation function, while v is a function representing the texture or noise. The algorithm proposed uses differential equations and is simple to solve. It also explains that the method can be used for discrimination of textures and texture segmentation.

In [46], the author gives a theory of Marr's primal sketch that integrates three components like texture model, a generative model along with image primitives, and a Gestalt field. It also defines the term "stretch ability," which aids in the division of images into texture and geometry, and it examines two different types of models—namely the detailed Markov random field model and the generative wavelet/sparse coding model. In this method, texture data retrieved from complex degraded images was retrieved and analysed by introducing an improved algorithm. Firstly, discrete wavelet transform (DWT) was used [47] to identify the edges present in images. Afterwards, the textual regions were localized by applying the Ada Boost classifier and CC clustering. Initially, [48] introduced morphological reconstructions that are based on techniques like geodesic transform which emphasize objects that were available in the centre of image and also remove the light and dark structures of concern to the border of image. This binarization process explained here is found to be much better than the other text binarization techniques.

3 PROBLEM STATEMENT AND MOTIVATION

The detection and recognition of text is considered as an active field of research in the numerous research communities concentrated on pattern recognition and computer vision. The recent development of various portable device– and smartphone-based applications may consider text detection and recognition as an active research area. Nowadays, detection of text could also be considered a challenging and more complicated task because there is a transparent difference between separating textual or non-textual regions while separating each character from the context. This makes the automated extraction of text much harder. Moreover, the most important reason that makes the identification of text and recognition in the natural scenes difficult is the intensity of illumination. A photograph's illumination is additionally affected by available shadows and lighting conditions of the environment, but the complex backgrounds of the image are normally obtained from outdoor images, which makes automated text extraction process harder. As a result, proper filtering must be used for text detection. They normally retrieve subimages from the primary image that they estimate as textual or non-textual. They also repeat this with sliding windows of varied scales. To eliminate this repetitive process, an efficient classifier that identifies the textual and non-textual part from the natural scene images with less error classification is required.

4 PROPOSED METHODOLOGY

Table 10.1 shows the algorithm for suggested methodology, which includes the steps for text extraction from complex degraded images.

4.1 EVALUATION METRICS

Few of the known text extraction techniques that are taken for the purpose of comparison are Ansari, He, Almaz'an [49–51], Khlif [18], Zhu, Zhang, R-FCN

TABLE 10.1
Algorithm for entire text extraction process

Input: Natural scene image with text
Output: Extracted text
Step 1: Preprocess the input image using guided image filter (GIF) for contrast enhancement.
Step 2: Segment the contrast-enhanced image using a marker-based watershed segmentation algorithm.
Step 3: Features present in segmented image are retrieved by using Gabor transform (GT) and stroke width transform (SWT).
Step 4: Based on these extracted features, the weighted Naïve Bayes classifier (WNBC) identifies the textual and non-textual parts.
Step 5: The errors occurring during the process of classification are minimized by using an EPO algorithm by providing an optimal solution, and it also prevents the solution from falling into the local optimum.
Step 6: The classified textual part is then given for a deep neural network for character recognition.
Step 7: Optimal parameter (weight) selection is necessary for DNN, which is achieved by a Adaptive Galactic Swarm Optimization (AGSO) algorithm.
Step 8: The classification errors that occur during text extraction are mimimized by determining the Manhattan distance between the strings.
Step 9: Perform Lexicon search, if Manhattan distance is 0, the text or string is same, or else if the distance is 1, then the optimized word is obtained.

An Efficient Deep Neural Network

(Region-based Fully Convolutional Networks), Faster R-CNN [52–55]. The result metrics are obtained using four parameters they are True Positive (TP), True Negative (TN), False Positive (FP), and False Negative (FN).

- The text part that is correctly identified as text is determined by TP.
- The text part that is incorrectly identified as non-text part is determined by FN.
- The non-text part that is correctly identified as non-text is determined by TN.
- The non-text part that is incorrectly identified as text is determined by FP.

(a) Accuracy: This is a very important parameter for comparison.

$$Accuracy = \frac{TP + TN}{TP + TN + FP + FN} \quad (10.1)$$

The accuracy value of the proposed method is higher than the existing method of text extraction as shown in Table 10.2. Few of the known approaches like Ansari, He, and Almaz'an [49–51] are taken as the existing method. Based on these values, the following graph is plotted. The accuracy of this work is found to be higher than the remaining three existing techniques as shown in Figure 10.1.

TABLE 10.2
Accuracy of proposed and prevailing methods

Methods	Accuracy (%)
Proposed	98.39
Ansari et al. [49]	95.3
He et al. [50]	94
Almaz'an et al. [51]	88.6

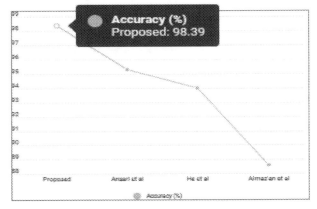

FIGURE 10.1 Accuracy (%) of proposed and existing methods.

(b) Precision: The ratio of true positives (TP) with overall detections is known as precision. The comparison of the precision of the suggested method with the remaining existing methods is shown in Table 10.3 and Figure 10.2. Precision can be expressed in the mathematical equation as shown in Equation (10.2),

$$p = \frac{TP}{TP + FP} \tag{10.2}$$

(c) Recall: The ratio between the identified true text and entire TP is determined by recall. The recall metrics can be expressed mathematically by equation as shown

TABLE 10.3
Precision of proposed and prevailing methods

Methods	Precision (%)
Proposed	93.79
Khlif [18]	89.94
Zhu [52]	83
Zhang [53]	78
R-FCN [54]	90
Faster R-CNN [55]	86

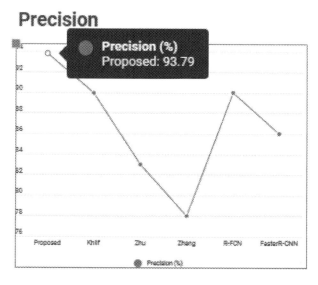

FIGURE 10.2 Precision (%) of proposed and existing methods.

An Efficient Deep Neural Network

in (10.3). The comparison of the precision of the suggested method with the remaining existing methods is shown in Table 10.4 and Figure 10.3.

$$r = \frac{TP}{TP + FN} \tag{10.3}$$

(d) F1-score: It is a very important parameter among the various performance parameters. It is used as indicator for proposed method. In F1-score evaluation, both parameters, precision and recall, are used. The F1-score is can be expressed in the mathematical equation as shown in (10.4). The comparison of the precision of the

TABLE 10.4
Recall of proposed and prevailing methods

Methods	Recall (%)
Proposed	96.8
Khlif [18]	82.28
Zhu [52]	84
Zhang [53]	88
R-FCN [54]	76
Faster R-CNN [55]	75

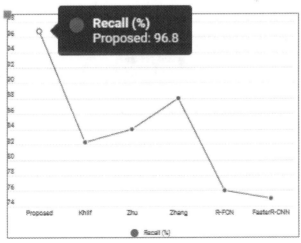

FIGURE 10.3 Recall (%) of proposed and existing methods.

suggested method with the remaining existing methods is shown in Table 10.5 and Figure 10.4.

$$F_1 = 2 \times \frac{p \times r}{p+r} \qquad (10.4)$$

4.2 Comparative Analysis

The precision, recall, and F1-score performance parameter values of proposed and existing text extraction techniques are shown in Figure 10.5. The performance

TABLE 10.5
F1-score of proposed and prevailing methods

Methods	F1-Score (%)
Proposed	95.2
Khlif [18]	85.94
Zhu [52]	84
Zhang [53]	83
R-FCN [54]	83
Faster R-CNN [55]	80

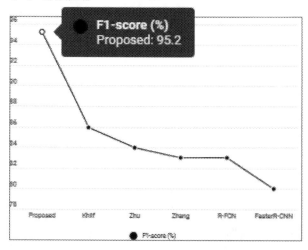

FIGURE 10.4 F1-score (%) of proposed and existing methods.

An Efficient Deep Neural Network 133

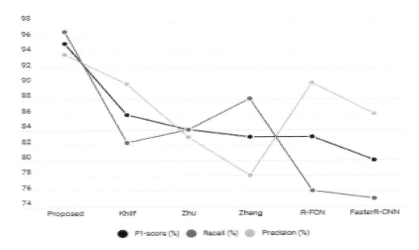

FIGURE 10.5 The comparative analysis precision, recall, and F1-score.

metrics results values are expressed as percentages. The available values indicate that this proposed method performs better text extraction from natural complex degraded images than the other existing techniques.

5 CONCLUSION

In this work, a deep learning (DL) –based optimization approach is deployed for text extraction from complex degraded images. First, we used GIF as a preprocessing step for text extraction, then a marker-watershed segmentation, SWT, GT, and WNBC techniques for segmentation, features extraction, and detection of textual and non-textual parts from a complex degraded image. Finally, a combination of DNN and AGSO is used for recognizing classified text parts. Dataset IIIT5K is used for the implementation part, and high performance is obtained with a parameter like accuracy, recall, precision, and F1-score, but occasionally, the characters may stray; alternatively, the similar character is often extracted several times, which can cause the extraction of incorrect textual data from the natural images. Therefore, in the future, an efficient technique must be implemented to avoid such defects in the text retrieval process. The performance of neural network architecture is enhanced further by using an optimization technique.

REFERENCES

[1] Yin, X.C., Yin, X., Huang, K., and Hao, H.W.: 'Robust text detection in natural scene images', *IEEE Transactions on Pattern Analysis and Machine Intelligence*, 2013, 36, (5), pp. 970–983.

[2] Wang, L., Uchida, S., Zhu, A., and Sun, J.: 'Human reading knowledge inspired text line extraction', *Cognitive Computation*, 2018, 10, (1), pp. 84–93.

[3] Wang, Y., Shi, C., Xiao, B., Wang, C., and Qi, C.: 'CRF based text detection for natural scene images using convolutional neural network and context information', *Neurocomputing*, 2018, 295, pp. 46–58.

[4] Paul, S., Saha, S., Basu, S., Saha, P.K., and Nasipuri, M.: 'Text localization in camera captured images using fuzzy distance transform based adaptive stroke filter', *Multimedia Tools and Applications*, 2019, pp. 1–20.

[5] Ma, J., Shao, W., Ye, H., Wang, L., Wang, H., Zheng, Y., and Xue, X.: 'Arbitrary-oriented scene text detection via rotation proposals', *IEEE Transactions on Multimedia*, 2018, 20, (11), pp. 3111–3122.

[6] Ghai, D., and Jain, N.: 'Comparative analysis of multi-scale Wavelet decomposition and k-means clustering based text extraction', *Wireless Personal Communications*, 2019, pp. 1–36.

[7] Pandey, D., Pandey, B.K., and Wairya, S.: 'Hybrid deep neural network with adaptive galactic swarm optimization for text extraction from scene images', *Soft Comput*, 2020. https://doi.org/10.1007/s00500-020-05245-4

[8] Dutta, I.N., Chakraborty, N., Mollah, A.F., Basu, S., and Sarkar, R.: 'Multi-lingual text localization from camera captured images based on foreground homogeneity analysis', In *Recent Developments in Machine Learning and Data Analytics*, Springer Nature, Switzerland AG, 2019, pp. 149–158.

[9] Tian, C., Xia, Y., Zhang, X., and Gao, X.: 'Natural scene text detection with MC – MR candidate extraction and coarse-to-fine filtering', *Neurocomputing*, 2017, April, 260, pp. 112–122.

[10] Ahmed, S.B., Naz, S., Razzak, M.I., and Yusof, R.B.: 'A novel dataset for English-Arabic Scene Text Recognition (EASTR)-42K and its evaluation using invariant feature extraction on detected extremal regions', *IEEE Access*, 2019, 7, pp. 19801–19820.

[11] Khare, V., Shivakumara, P., Raveendran, P., and Blumenstein, M.: 'A blind deconvolution model for scene text detection and recognition in video', *Pattern Recognition*, 2016, 54, pp. 128–148.

[12] Ahmed, S.B., Naz, S., Razzak, M.I., Rashid, S.F., Afzal, M.Z., and Breuel, T.M.: 'Evaluation of cursive and non-cursive scripts using recurrent neural networks', *Neural Computing and Applications*, 2016, 27, (3), pp. 603–613.

[13] Francis, L.M., and Sreenath, N.: 'Robust scene text recognition: Using manifold regularized Twin-Support Vector Machine', *Journal of King Saud University-Computer and Information Sciences*, 2019. In Press.

[14] Mehmood, Z., Mahmood, T., and Javid, M.A.: 'Content-based image retrieval and semantic automatic image annotation based on the weighted average of triangular histograms using support vector machine', *Applied Intelligence*, 2018, 48, (1), pp. 166–181.

[15] Yin, X.C., Pei, W.Y., Zhang, J., and Hao, H.W.: 'Multi-orientation scene text detection with adaptive clustering', *IEEE Transactions on Pattern Analysis and Machine Intelligence*, 2015, 37, (9), pp. 1930–1937.

[16] Wu, H., Zou, B., Zhao, Y.Q., and Guo, J.: 'Scene text detection using adaptive color reduction, adjacent character model and hybrid verification strategy', *The Visual Computer*, 2017, 33, (1), pp. 113–126.

[17] Sain, A., Bhunia, A.K., Roy, P.P., and Pal, U.: 'Multi-oriented text detection and verification in video frames and scene images', *Neurocomputing*, 2018, 275, pp. 1531–1549.

[18] Khlif, W., Nayef, N., Burie, J.C., Ogier, J.M., and Alimi, A.: 'Learning text component features via convolutional neural networks for scene text detection', In *2018 13th IAPR International Workshop on Docu.nent Analysis Systems (DAS)*, IEEE, Vienna, Austria, 2018, April, pp. 79–84.

[19] Tang, Y., and Wu, X.: 'Scene text detection using superpixel-based stroke feature transform and deep learning based region classification', *IEEE Transactions on Multimedia*, 2018, 20, (9), pp. 2276–2288.

[20] Bhunia, A.K., Kumar, G., Roy, P.P., Balasubramanian, R., and Pal, U.: 'Text recognition in scene image and video frame using Color Channel selection', *Multimedia Tools and Applications*, 2018, 77, (7), pp. 8551–8578.

[21] Ali, A., Pickering, M., and Shafi, K.: 'Urdu natural scene character recognition using convolutional neural networks', In *2018 IEEE 2nd International Workshop on Arabic and Derived Script Analysis and Recognition (ASAR)*, IEEE, London, 2018, March, pp. 29–34.

[22] Huang, Z., Zhong, Z., Sun, L., and Huo, Q.: 'Mask R-CNN with pyramid attention network for scene text detection', In *2019 IEEE Winter Conference on Applications of Computer Vision (WACV)*, IEEE, Waikoloa Village, HI, 2019, January, pp. 764–772.

[23] Baran, R., Partila, P., and Wilk, R.: 'Automated text detection and character recognition in natural scenes based on local image features and contour processing techniques', In *International Conference on Intelligent Human Systems Integration*, Springer, Cham, 2018, January, pp. 42–48.

[24] Xue, M., Shivakumara, P., Zhang, C., Lu, T., and Pal, U.: 'Curved text detection in blurred/non-blurred video/scene images', *Multimedia Tools and Applications*, 2019, pp. 1–25.

[25] Wang, Y., Wang, L., and Su, F.: 'A robust approach for scene text detection and tracking in video', In *Pacific Rim Conference on Multimedia*, Springer, Cham, 2018, September, pp. 303–314.

[26] Antonini, M., and Barlaud, M.: 'Image coding using wavelet transform', *IEEE Transactions on Image Processing*, April 1992, 1, (2).

[27] Duchowski, A.T.: 'Representing multiple region of interest with Wavelets', In *Proc. SPIE 3309, Visual Communications and Image Processing '98*, San Jose, CA, January 9, 1998, 975.

[28] Duchowski, A.T.: 'Representing multiple region of interest images with Wavelets', In *Proc. SPIE 1109, Visual Communications and Image Processing '99*, Macau, China, April 10, 1999, 960.

[29] Antonini, M., and Barlaud, M.: 'Image coding using wavelet transform', *IEEE Transactions on Image Processing*, 1, (2), April 1992.

[30] Chen, C.T., and Chen, L.G.: 'A self-adjusting weighted median filter for removing impulse noise in images', In *Image Processing, 1996. Proceedings., International Conference on* (Vol. 1). IEEE, Lausanne, Switzerland,1996, September, pp. 419–422.

[31] Peng, S.F., and Lucke, L.: 'A hybrid filter for image enhancement', In *icip*. IEEE, Washington, DC, 1995, October, p. 163.

[32] Sauvola, J., and Pietikäinen, M.: 'Adaptive document image binarization', *Pattern Recognition*, 2000, 33, (2), pp. 225–236.

[33] Sobottka, K., Kronenberg, H., Perroud, T., and Bunke, H.: 'Text extraction from colored book and journal covers', *International Journal on Document Analysis and Recognition*, 2000, 2, (4), pp. 163–176.

[34] Yuan, Q., and Tan, C.L.: 'Text extraction from gray scale document images using edge information', In *Document Analysis and Recognition, 2001. Proceedings. Sixth International Conference on*. IEEE, Seattle, Washington, DC, 2001, pp. 302–306.

[35] Tsai, C.M., and Lee, H.J.: 'Binarization of color document images via luminance and saturation colorfeatures', *Image Processing, IEEE Transactions on*, 2002, 11, (4), pp. 434–451.

[36] Gllavata, J., Ewerth, R., and Freisleben, B.: 'A robust algorithm for text detection in images', In *Image and Signal Processing and Analysis, 2003. ISPA 2003. Proceedings of the 3rd International Symposium on* (Vol. 2). IEEE, 2003, September, pp. 611–616.

[37] Kim, K.I., Jung, K., and Kim, J.H.: 'Texture-based approach for text detection in images using support vector machines and continuously adaptive mean shift algorithm', *Pattern Analysis and Machine Intelligence, IEEE Transactions on*, 2003, 25, (12), pp. 1631–1639.

[38] Jung, K., Kim, K.I., and Jain, A.K.: 'Text information extraction in images and video: A survey', *Pattern Recognition*, 2004, 37, (5), pp. 977–997.

[39] Ye, Q., Huang, Q., Gao, W., and Zhao, D.: 'Fast and robust text detection in images and video frames', *Image and Vision Computing*, 2005, 23, (6), pp. 565–576.

[40] Chan, R.H., Ho, C.W., and Nikolova, M.: 'Salt-and-pepper noise removal by median-type noise detectors and detail-preserving regularization', *Image Processing, IEEE Transactions on*, 2005, 14, (10), pp. 1479–1485.

[41] Shui, P.L.: 'Image denoising algorithm via doubly local Wiener filtering with directional windows in wavelet domain', *Signal Processing Letters, IEEE*, 2005, 12, (10), pp. 681–684.

[42] Gatos, B., Pratikakis, I., and Perantonis, S.J.: 'Adaptive degraded document image binarization', *Pattern Recognition*, 2006, 39, (3), pp. 317–327.

[43] Starck, J.L., Elad, M., and Donoho, M.: 'Redundant multiscale transforms and their application for morphological component separation', *Advances in Imaging and Electron Physics*, 2004, 132.

[44] Starck, J.L., Elad, M., and Donoho, D.L.: 'Image decomposition via the combination of sparse representations and a variational approach', *IEEE Trans on Image Processing*, October 2005, 14, pp. 1570–1582.

[45] Vese, L., and Osher, S.: 'Modeling textures with total variation minimization and oscillating pattern in image processing', *Journal of Scientific Computing*, 2003, 19, pp. 553–577.

[46] Guo, C., Zhu, S., and Wu, Y.: 'Towards a mathematical theory of primal sketch and sketchability', In *Proceedings of the Ninth IEEE International Conference on Computer Vision (ICCV)*, Nice, France, 2003, October.

[47] Kumuda, T., and Basavaraj, L.:'Edge based segmentation approach to extract text from scene images', In *2017 IEEE 7th International Advance Computing Conference (IACC)*, IEEE, Hyderabad, India, 2017, January, pp. 706–710.

[48] Trémeau, A., Fernando, B., Karaoglu, S., and Muselet, D.: 'Detecting text in natural scenes based on a reduction of photometric effects: Problem of text detection', In *International Workshop on Computational Color Imaging*, Springer, Berlin, Heidelberg, 2011, April, pp. 230–244.

[49] Ansari, G.J., Shah, J.H., Yasmin, M., Sharif, M., and Fernandes, S.L.: 'A novel machine learning approach for scene text extraction', *Future Generation Computer Systems*, 2018, 87, pp. 328–340.

[50] He, P., Huang, W., Qiao, Y., Loy, C.C., and Tang, X.: 'Reading scene text in deep convolutional sequences', In *Thirtieth AAAI Conference on Artificial Intelligence*, Phoenix, AZ, 2016, March.

[51] Almazán, J., Gordo, A., Fornés, A., and Valveny, E.: 'Word spotting and recognition with embedded attributes', *IEEE Transactions on Pattern Analysis and Machine Intelligence*, 2014, 36, (12), pp. 2552–2566.

[52] Zhu, A., and Uchida, S.: 'Scene text relocation with guidance', In *2017 14th IAPR International Conference on Document Analysis and Recognition (ICDAR)*, IEEE, Kyoto, Japan, 2017, November, 1, pp. 1289–1294.

[53] Zhang, Z., Zhang, C., Shen, W., Yao, C., Liu, W., and Bai, X.: 'Multi-oriented text detection with fully convolutional networks', In *Proceedings of the IEEE Conference on Computer Vision and Pattern Recognition*, Las Vegas, NV, 2016, pp. 4159–4167.

[54] Dai, J., Li, Y., He, K., and Sun, J.: 'R-fcn: Object detection via region-based fully convolutional networks', In *Advances in Neural Information Processing Systems*, Barcelona, Spain, 2016, pp. 379–387.

[55] Ren, S., He, K., Girshick, R., and Sun, J.: 'Faster R-CNN: Towards real-time object detection with region proposal networks', In *Advances in Neural Information Processing Systems*, Montréal, Canada, 2015, pp. 91–99.

11 Diet Recommendation Model of Quality Nutrition for Cardiovascular Patients

Surbhi Vijh & Sanjay Kumar Dubey

CONTENTS

1 Introduction 139
2 Proposed Methodology..140
   2.1 Steps for Analytic Hierarchy Process..141
3 Experimental Work..142
   3.1 Validation Using Fuzzy TOPSIS 144
4 Results and Discussion 148
5 Conclusions..149
References..150

1 INTRODUCTION

Cardiovascular disease occurs when problems arise in the working of the blood vessels and heart. There are various causes of cardiovascular disease; among them, atheroma is most common. In atheroma, a fatty deposit builds up inside the walls of arteries. The fat deposits narrow the wall, minimizing the flowing capacity of blood. However, risk factors of cardiovascular disease are divided into three categories: lifestyle related-risk factors, treatable risk factors and permanent risk factors. Lifestyle-related risk factors are unbalanced diet, smoking, excess alcohol, excess salt intake, obesity and insufficient physical activity. Treatable risk factors are high blood pressure, diabetes, high cholesterol level and high fat blood level. Permanent risk factors are heredity, age and gender (Patient 2019). Some types of cardiovascular disease are arrhythmia, congenital heart disease, heart failure, coronary artery disease, stroke and heart attack. Arrhythmia is caused by irregular heartbeat, either too fast or too slow. Congenital heart disease occurs when a person is born with a defect in the heart, either its shape or its function. Heart attack happens due to blockage of blood flow in the heart. Coronary artery disease causes the deposition of cholesterol on the coronary artery walls. Heart failure arises due to insufficient oxygen or blood in the body for smooth functioning (Medical News Today 2019), (Heart 2019). Symptoms of cardiovascular disease are chest pain, chest discomfort, chest pressure, shortness of breath, numbness, pain in back or upper abdomen, abnormal heartbeat, weakness in arms or legs, dizziness, fatigue, swelling in feet or hands and dry cough (Mayo Clinic 2019). Chances of heart disease development in women is around 10 years later than men, but the death rate of women due to heart diseases is still more than that of men. However, a balanced diet is recommended to prevent and overcome the challenges of cardiovascular disease.

Ahmed and Mahmoud (2020) presented an expert system to help diabetes patients on the basis of diet recommendation. The meal plan is designed by medical experts and researchers and has undergone various stages in order to formalize the implantation. Diet recommendation for diabetes patients is designed using the type 2 fuzzy ontology method. An intelligent diet recommendation system (Lee et al. 2010) is developed for diabetes patients. Fuzzy logic and ontology are applied for study of diet recommendation in the fields of Ayurveda and prakriti. This approach provides recommendation of diet to patients with better results and certainty (Chavan & Sambare 2015). A rational diet recommendation system is constructed on the basis of various parameters such as kidney function, height, weight, activity levels, hypertension and hyperlipidemia using fuzzy rules and knapsack methods. The knapsack method is used to suggest or recommend the appropriate food items to the user (Chen et al. 2013). Moreover, integrated solutions for the patient can be achieved using an analytical hierarchy approach and fuzzy technique for order preferences by similarity to ideal solution method. The information is shared (Shukla et al. 2014) and validated based upon coordination criteria considered on the basis of expert suggestion and survey. The development of a healthy dietary pattern is implemented using fuzzy algorithms having all micronutrients and can be suggested to patients based on various factors such as age (Priyono & Surendro 2013).

In this research work, the analytical hierarchy process (Saaty 1980) is implemented for diet recommendation for cardiovascular patients. The proper meal is developed containing all important nutrients, and emphasis is given to providing the appropriate nutritional diet beneficial (Vijh et al. 2020). The formation of a diet plan is determined by considering morning, lunch, dinner (Saini & Dubey 2017) through suggestions from expert dietitians, researchers and practitioners. The alternative diets are presented, and the most suitable among them is selected based on criteria and ranking (Arvizu et al. 2020). Moreover, the validation of the proposed work for cardiovascular disease is performed using the fuzzy TOPSIS (Singh & Dubey 2017) method for determining the order of preferences. The consistency of the proposed work is checked by implementing the validation. The approach is preferred over other approaches, as it provides concrete details in comparison to conventional approaches. The consistency of proposed work is checked by implementing the validation in proposed work.

2 PROPOSED METHODOLOGY

AHP stands for analytic hierarchy process based on multicriteria decision-making analysis, or MCDM, technique applied to determine the decision upon some criteria for the particular application. The analytical hierarchy approach was formulated by Thomas L. Saaty in 1980. The critical factors are considered for solving

Diet Recommendation Model

TABLE 11.1
Comparison matrix

Y	Ci	Cj	Cn
Ci	1	Cij	Cin
Cj	1/Cij	1	Cjn
Cn	1/Cin	1/Cjn	1

TABLE 11.2
Scale for evaluation

N	1	2	3	4	5	6
C.N	0.0	0.0	0.58	0.90	1.12	1.24

TABLE 11.3
Consistency table

Elements	Significance
1	Equally valuable
3	Moderately valuable
5	Very valuable
7	Extremely valuable
9	Absolutely valuable
2,4,6,8	Intermediate valuable

complex problems using mathematical solutions and making the comparison among the alternatives. The selected nutrients are prioritized according to their importance using AHP. Eigenvalues and vectors, consistency index and ratio are calculated to showcase the result. The method evaluates the consistency of decisions, which removes biasness and errors in the decision making process. The comparison matrix, evaluation scale and consistency table are shown in Tables 11.1, 11.2 and 11.3, respectively.

2.1 STEPS FOR ANALYTIC HIERARCHY PROCESS

- Determine the critical factors and considerable alternative.
- Using Saaty's fundamental scale, create a pairwise comparison matrix.
- Normalize nth root of product and acquire desired assigned weights.
- Calculate consistency index by applying Saaty's consistency table.
- Calculate the consistency ratio and reliability index of the alternative.
- Computation of the final ranking hierarchy process.

3 EXPERIMENTAL WORK

The nutrition alternative diet chart is designed for cardiovascular patients based upon a few critical factors. The parameters are considered with the help of surveys, practitioners and experts. The diet chart consists of essential nutrition for the breakfast, lunch and dinner containing necessary meals as shown in Table 11.5. The proposed model using AHP is presented in Figure 11.1, showing the requirement and recommendation for cardiovascular disease. Tables 11.6, 11.7, 11.8 and 11.9 represent the attribute analysis with respect to diet A1, B1 and C1, respectively. Table 11.10 depicts the significance of elements. Table 11.11 shows the ranking and priority of alternative diets.

TABLE 11.4
Importance of attributes in diet

Attributes	Importance in Diet
Magnesium	In the pathogenesis of cardiovascular disease, magnesium has an important role at the biochemical and cellular levels. It helps in proper maintenance, functioning of cellular membrane and mitochondria; therefore; its deficiency can lead to mortality and morbidity (DiNicolantonio et al. 2018).
Vitamin C	L-ascorbic acid is an essential dietary component in antibody production. It benefits in a variety of ways. It is considered a strong antioxidant, which helps to protect against heart disease. The proper amount of fruit and vegetable intake may help to reduce the risk of cardiovascular disease. It prevents atherosclerosis (Hercberg et al. 1998).
Omega 3	It helps in improving lung functionality. It plays a role in improving flow-mediated arterial dilation and influences heart rate (Mohebi-Nejad & Bikdeli 2014).
Coenzyme 10	The supplement of CQ10 prevents disease by maintaining optimal cellular and mitochondrial function (Langsjoen & Langsjoen 1999).

TABLE 11.5
Diet components

Diet List	Attributes
Diet 1 (A1)	Breakfast: 1 egg white omelet, avocado, raisin
	Lunch: 1 whole grain bread, spinach, soybeans
	Dinner: spinach soup, roti, turnip greens and daal
Diet 2 (B1)	Breakfast: sandwich, milk, fruit salad
	Lunch: whole-grain bread, spinach, soybeans
	Dinner: brown rice, chickpeas, salad
Diet 3 (C1)	Breakfast: whole-grain bread sandwich with peanut butter, orange juice, walnuts
	Lunch: roti, daal, cauliflower, curd
	Dinner: steamed vegetables like broccoli, beans, carrot and spinach, curd

Diet Recommendation Model

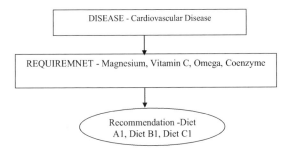

FIGURE 11.1 AHP model.

TABLE 11.6
Magnesium with respect to A1, B1, C1

λavg (max) = 3.1333, C.I. = 0.0666 C.R. = 0.1149

Magnesium	A1	B1	C1	Eigenvector (w)
A1	1	0.5	0.33	0.1672
B1	2	1	2	0.4839
C1	3	0.5	1	0.3489

TABLE 11.7
Vitamin C with respect to A1, B1, C1

λavg(max) = 3.0537, C.I. = 0.0268 C.R. = 0.0463

Vitamin C	A1	B1	C1	Eigenvector (w)
A1	1	0.14	0.2	0.0715
B1	7	1	3	0.6500
C1	5	0.33	1	0.2784

TABLE 11.8
Omega with respect to A1, B1, C1

λavg(max) = 3.0940, C.I. = 0.0470 C.R. = 0.0810

Omega	A1	B1	C1	Eigenvector (w)
A1	1	0.5	0.25	0.1265
B1	2	1	0.2	0.1865
C1	4	5	1	0.6870

TABLE 11.9
Coenzyme with respect to A1, B1, C1

λavg(max) = 3.4241, C.I. = 0.2120, C.R. = 0.3656

Coenzyme10	A1	B1	C1	Eigenvector (w)
A1	1	0.33	0.14	0.0954
B1	3	1	3	0.5531
C1	7	0.33	1	0.3515

TABLE 11.10
Elements with respect to their significance
$\lambda avg(max) = 4.0779$, C.I. = 0.0260, C.R. = 0.0289

Element	Magnesium	Vitamin C	Omega	Coenzyme 10	Eigenvector (w)
Magnesium	1	1	0.2	0.143	0.0733
Vitamin C	1	1	0.33	0.2	0.0903
Omega	5	3	1	0.33	0.2658
Coenzyme	7	5	3	1	0.5705

TABLE 11.11
Ranking table

	Magnesium	Vitamin C	Omega	Coenzyme 10	Priority	Outcome
A1	0.1672	0.0715	0.1265	0.0954	0.1068	3
B1	0.4839	0.6500	0.1865	0.5531	0.4593	1
C1	0.3489	0.2784	0.6870	0.3515	0.4339	2

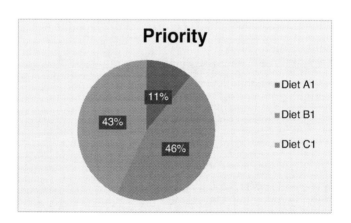

FIGURE 11.2 Ranking graph for alternative diet.

3.1 Validation Using Fuzzy TOPSIS

The fuzzy TOPSIS method is utilized for validating the result of AHP. The methodology was developed by C. Cheng and Hwang (Chen 2000). Fuzzy TOPSIS finds the nearest affirmative solution depending upon the closeness coefficient, referred to as level of closeness of diet to the affirmative solution. The fuzzy TOPSIS technique eliminates the uncertainty about the category of diet by robustly proposing the special varieties of meals that can be eaten by the cardiovascular patient. Tables 11.12 and 11.13 show the linguistic variable for fitness criteria.

Diet Recommendation Model

TABLE 11.12
Linguistic variable showing fitness criteria

Very Poor (VP)	(0, 0, 1)
Poor (P)	(0, 1, 3)
Medium Poor (MP)	(1, 3, 5)
Medium (M)	(3, 5, 7)
Medium Good (MG)	(5, 7, 9)
Good (G)	(7, 9, 10)
Very Good (VG)	(9, 10, 10)

TABLE 11.13
Linguistic variable for fitness criteria

Very Low (VL)	(0, 0, 0.1)
Low (L)	(0, 0.1, 0.3)
Medium Low (ML)	(0.1, 0.3, 0.5)
Medium (M)	(0.3, 0.5, 0.7)
Medium High (MH)	(0.5, 0.7, 0.9)
High (H)	(0.7, 0.9, 1.0)
Very High (VH)	(0.9, 1.0, 1.0)

The mentioned steps are performed for applying the Fuzzy TOPSIS technique:

Step 1: Initially, a decision table for the diet recommendation system is constructed as shown in Table 11.14.

Step 2: Basically, three doctors and experts are consulted for making the diet plan. Experts suggest a complete-day diet chart including breakfast, lunch and dinner. The research and observation show that the accurate amount of nutrients, i.e. vitamin C, omega 3, CQ10 and magnesium, are required to be present in a cardiovascular patient's diet as shown in Table 11.15.

Step 3: Conversion of doctor suggestion into fuzzy numbers for all the three diets is presented in Table 11.16, i.e. diet A1, diet B1 and diet C1. Each diet includes three meals per day.

Step 4: Weights of all the four factors have been evaluated.

Step 5: Normalized fuzzy matrix has been constructed by dividing each number in the matrix by the highest number in that column in Table 11.17.

Step 6: Weight matrix and normalized fuzzy matrix have been multiplied as shown in Table 11.18.

Step 7: Next, evaluate the fuzzy positive ideal solution (FPIS) and fuzzy negative ideal solution (FNIS) as represented in $A^{\sim *}$ and $A^{\sim}$ through Tables 11.19 and 11.20.

$A^{\sim *}= [(1, 1, 1), (1, 1, 1), (1, 1, 1), (1, 1, 1)]$
$A^{\sim}-= [(0, 0, 0), (0, 0, 0), (0, 0, 0), (0, 0, 0)]$

Step 8: Evaluate distance of each diet from FPIS and FNIS.

Step 9: Determine the value of closeness coefficient.
Step 10: Final ranking table is obtained with the help of closeness coefficient as shown in Table 11.21.

The mechanism of the fuzzy TOPSIS technique is represented in the following steps, and required implementation is performed for validation purposes.

Step 1

TABLE 11.14
Decision table

Factors	Diet A1	Diet B1	Diet C1
Magnesium	M	MH	MH
Vitamin C	H	H	MH
Omega	ML	ML	M
CQ10	M	VL	ML

Step 2

TABLE 11.15
Decision-maker table

Criteria		System		Decision-Makers	
	DIET A1	G		VG	G
Magnesium	DIET B1	G		MG	G
	DIET C1	F		MG	F
	DIET A1	MG		G	MG
Vitamin C	DIET B1	MG		F	F
	DIET C1	F		MP	F
	DIET A1	G		MG	G
Omega	DIET B1	MG		MG	MG
	DIET C1	P		MP	P
	DIET A1	MG		G	G
CO10	DIET B1	F		F	MP
	DIET C1	P		MP	MP

Step 3

$X11 = 1/3 [(7,9,10) + (9,10,10) + (7,9,10)] = (7.66, 9.33, 10)$
$X21 = 1/3 [(7,9,10) + (5,7,9) + (7,9,10)] = (6.33, 8.33, 9.66)$
$X31 = 1/3 [(3,5,7) + (5,7,9) + (3,5,7)] = (3.66, 5.66, 7.66)$
$X12 = 1/3 [(5,7,9) + (7,9,10) + (5,7,9)] = (5.66, 7.66, 9.33)$
$X22 = 1/3 [(5,7,9) + (3,5,7) + (3,5,7)] = (3.66, 5.66, 7.66)$
$X32 = 1/3 [(3,5,7) + (1,3,5) + (3,5,7)] = (2.33, 4.33, 6.33)$
$X13 = 1/3 [(7,9,10) + (5,7,9) + (7,9,10)] = (6.33, 8.33, 9.66)$
$X23 = 1/3 [(5,7,9) + (5,7,9) + (5,7,9)] = (5,7,9)$

Diet Recommendation Model

X33 = 1/3 [(0,1,3) + (1,3,5) + (0,1,3)] = (0.33,1.66,3.66)
X14 = 1/3 [(5,7,9) + (7,9,10) + (7,9,10)] = (6.33,8.33,9.66)
X24 = 1/3 [(3,5,7) + (3,5,7) + (1,3,5)] = (2.33,4.33,6.33)
X34 = 1/3 [(0,1,3) + (1,3,5) + (1,3,5)] = (0.66,2.33,4.33)

TABLE 11.16
Fuzzy numbers

(7.66, 9.33, 10)	(5.66, 7.66, 9.33)	(6.33, 8.33, 9.66)	(6.33, 8.33, 9.66)
(6.33, 8.33, 9.66)	(3.66, 5.66, 7.66)	(5,7,9)	(2.33, 4.33, 6.33)
(3.66, 5.66, 7.66)	(2.33, 4.33, 6.33)	(0.33, 1.66, 3.66)	(0.66, 2.33, 4.33)

Step 4

W1 = 1/3[(0.3,0.5,0.7) + (0.5,0.7,0.9) + (0.5,0.7,0.9)] = (0.43,0.63,0.83)
W2 = 1/3[(0.7,0.9,1.0) + (0.7,0.9,1.0) + (0.5,0.7,0.9)] = (0.63,0.83,0.96)
W3 = 1/3[(0.1,0.3,0.5) + (0.1,0.3,0.5) +(0.3,0.5,0.7)] = (0.16,0.36,0.56)
W4 = 1/3[(0.3,0.5,0.7) + (0,0,0.1) + (0.1,0.3,0.5)] = (0.13,0.26,0.43)

Step 5

TABLE 11.17
Normalized fuzzy table

(0.766, 0.933, 1)	(0.61, 0.82, 1)	(0.66, 0.86, 1)	(0.66, 0.86, 1)
(0.633, 0.833, 0.966)	(0.39, 0.61, 0.82)	(0.52, 0.72, 0.93)	(0.24, 0.45, 0.66)
(0.366, 0.566, 0.766)	(0.25, 0.46, 0.68)	(0.03, 0.17, 0.38)	(0.07, 0.24, 0.45)

Step 6

TABLE 11.18
Weighted fuzzy number

(0.33, 0.59, 0.83)	(0.38, 0.68, 0.96)	(0.11, 0.31, 0.56)	(0.09, 0.22, 0.43)
(0.27, 0.52, 0.80)	(0.25, 0.51, 0.79)	(0.08, 0.26, 0.52)	(0.03, 0.12, 0.28)
(0.16, 0.36, 0.64)	(0.16, 0.38, 0.65)	(0.01, 0.06, 0.21)	(0.01, 0.06, 0.19)

Step 7

TABLE 11.19
Distance of each diet from A*

D A1*	2.67
D B1*	1.92
D C1*	1.78

Step 8

TABLE 11.20
Distance of each diet from $A1^-$

$DA1^-$	1.55
$DB1^-$	1.54
$DC1^-$	1.09

Step 9

TABLE 11.21
Closeness coefficient and ranking of diet

Diets	Closeness Coefficient	Ranking
Diet A1	0.367	3
Diet B1	0.446	1
Diet C1	0.379	2

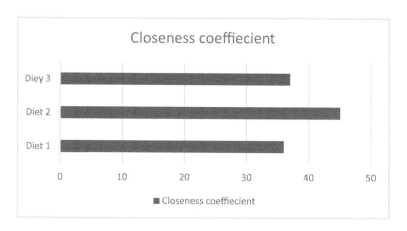

FIGURE 11.3 Closeness coefficient.

4 RESULTS AND DISCUSSION

In this chapter, the AHP technique is utilized to select the best edibles among the alternatives diet provided by doctor. The selected diet contains all the nutrients necessary to cure the cardiovascular diseases. The four most important nutrients – vitamin C, omega 3, CQ10 and magnesium – are chosen as shown in Table 11.4. Using AHP, it is observed that diet B1 is best among all considered diets. Diet B1 contains all the needed nutrients in the proper amounts. Table 11.11 shows the result analysis obtained by applying the AHP method, and Figure 11.2 depicts the ranking of alternatives. Further, the results acquired from AHP are validated using the

Diet Recommendation Model

TABLE 11.22
Comparison

DIET	AHP	Fuzzy TOPSIS	Rank
Diet A1	0.1068	0.367	3
Diet B1	0.4593	0.446	1
Diet C1	0.4339	0.379	2

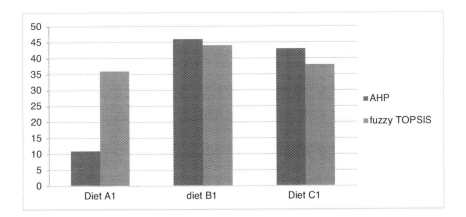

FIGURE 11.4 Final observation.

fuzzy TOPSIS technique. Table 11.19 represents the validated results, and Figure 11.3 shows the closeness coefficient obtained from the fuzzy TOPSIS method. So diet B1 is recommended over other diets. Hence, it is validated that the result of AHP is correct and diet B1 is preferred. The comparative analysis of results is shown in Table 11.22. Diet B1 has obtained highest ranking and A1 has obtained least rank as depicted in Figure 11.4.

5 CONCLUSIONS

Cardiovascular patients can suffer from a collection of diseases such as coronary heart problems, stroke, peripheral arterial disease and aortic disease. The proposed work is preceded by concentrating on an Indian perspective, since the diet considered is easily available in the Indian economy. With conventional techniques, the essential-nutrition meals are not presented properly; rather the overview is considered. However, in this chapter, the essential diet is evaluated among the alternatives, which is highly beneficial for patients. The AHP methodology is used to determine critical factors and their order of preference. The diet plan recommended in this chapter is designed based on the suggestions of expert doctors and researchers. Ranking among alternative diets is achieved. The experimental results are validated using the fuzzy TOPSIS method for the decision-making process to check the consistency of the proposed model and reduce the linguistic errors. Future research work includes more alternative nutritional diets, which can be converted into a mobile-based application upon more reliable findings.

REFERENCES

JOURNAL ARTICLES

Ahmed, I. M., & Mahmoud, A. M. (2020). Development of an expert system for diabetic type-2 diet. *arXiv preprint arXiv:2003.05104*.

Arvizu, M., Bjerregaard, A. A., Madsen, M. T., Granström, C., Halldorsson, T. I., Olsen, S. F., . . . Chavarro, J. E. (2020). Sodium intake during pregnancy, but not other diet recommendations aimed at preventing cardiovascular disease, is positively related to risk of hypertensive disorders of pregnancy. *The Journal of Nutrition*, *150*(1), 159–166.

Chavan, S. V., & Sambare, S. (2015). Study of diet recommendation system based on fuzzy logic and ontology. *International Journal of Computer Applications*, *132*(12), 20–24.

Chen, C. T. (2000). Extensions of the TOPSIS for group decision-making under fuzzy environment. *Fuzzy Sets and Systems*, *114*(1), 1–9.

DiNicolantonio, J. J., Liu, J., & O'Keefe, J. H. (2018). Magnesium for the prevention and treatment of cardiovascular disease.

Hercberg, S., Galan, P., Preziosi, P., Alfarez, M. J., & Vazquez, C. (1998). The potential role of antioxidant vitamins in preventing cardiovascular diseases and cancers. *Nutrition*, *14*(6), 513–520.

Langsjoen, P. H., & Langsjoen, A. M. (1999). Overview of the use of CoQ_{10} in cardiovascular disease. *Biofactors*, *9*(2–4), 273–284.

Lee, C. S., Wang, M. H., & Hagras, H. (2010). A type-2 fuzzy ontology and its application to personal diabetic-diet recommendation. *IEEE Transactions on Fuzzy Systems*, *18*(2), 374–395.

Mohebi-Nejad, A., & Bikdeli, B. (2014). Omega-3 supplements and cardiovascular diseases. *Tanaffos*, *13*(1), 6.

Priyono, R. A., & Surendro, K. (2013). Nutritional needs recommendation based on fuzzy logic. *Procedia Technology*, *11*, 1244–1251.

Saaty, T. L. (1980). The analytical hierarchy process, planning, priority. *Resource Allocation*. RWS publications, USA.

Saini, S., & Dubey, S. K. (2017). Recommendation of diet to jaundice patient on the basis of nutrients using AHP and fuzzy AHP technique. *International Journal of Intelligent Engineering and Systems*, *10*(4).

Shukla, R. K., Garg, D., & Agarwal, A. (2014). An integrated approach of Fuzzy AHP and Fuzzy TOPSIS in modeling supply chain coordination. *Production & Manufacturing Research*, *2*(1), 415–437.

Singh, M. P., & Dubey, S. K. (2017). Recommendation of diet to anaemia patient on the basis of nutrients using AHP and fuzzy TOPSIS approach. *International Journal of Intelligent Engineering and Systems*, *10*(4).

CONFERENCE PROCEEDINGS

Chen, R. C., Lin, Y. D., Tsai, C. M., & Jiang, H. (2013, June). Constructing a diet recommendation system based on fuzzy rules and knapsack method. In *International Conference on Industrial, Engineering and Other Applications of Applied Intelligent Systems* (pp. 490–500). Springer, Berlin, Heidelberg.

Vijh, S., Gaur, D., & Kumar, S. (2020, January). Diet recommendation for hypertension patient on basis of nutrient using AHP and entropy. In *2020 10th International Conference on Cloud Computing, Data Science & Engineering (Confluence), India* (pp. 364–368). IEEE.

Websites

Heart (2019, Sept 25). www.heart.org/en/health-topics/consumer-healthcare/what-is-cardiovascular-disease

Mayo Clinic (2019, Sept 30). www.mayoclinic.org/diseases-conditions/heart-disease/symptoms-causes/syc-20353118

Medical News Today (2019, Sept 22). www.medicalnewstoday.com/articles/257484.php

Patient (2019, Sept 22). https://patient.info/health/cardiovascular-disease-atheroma

12 Dynamic Simulation Model to Improve Travel Time Efficiency (Case Study: Surabaya City)

Shabrina Luthfiani Khanza, Erma Suryani, & Rully Agus Hendrawan

CONTENTS

1 Introduction 153
2 Literature Review 154
    a   System Dynamics Simulation to Improve Travel Time Efficiency 154
    b   Scenario Planning to Reduce Traffic Congestion 156
3 Model Development 156
4 Model Validation 157
5 Scenario Development 158
    5.1   Bus Rapid Transit Development Scenario 159
    5.2   Tram Development without Highway Expansion Scenario 160
    5.3   Tram Development with Highway Expansion Scenario 161
6 Conclusion and Further Research 162
Notes 162

1 INTRODUCTION

Traffic congestion is quite concerning problem, especially in big cities. As the second-most populated city in Indonesia, Surabaya also has heavy traffic in relation to private vehicle growth. The number of private vehicles in Surabaya always increases every year. In 2014, there were 2.3 million units of motorcycles, and it increased to 2.4 million units in 2015.[1] The private car, on average, increased by 7% annually from 1.9 million units in 2015 to 2 million units in 2016 and 2.2 million units in 2017.[2] The increase in the number of private vehicles will certainly lead to congestion. The number of vehicles is also increasing in Urip Sumoharjo street, one of the arterial roads in Surabaya that provide access to the downtown and connect two other arterial roads. Heavy traffic will increase travel time. Urip Sumoharjo, at 500 meters long needs less than a minute to traverse, but it can take up to 2 minutes or more in high traffic jams. This short delay can cause a bigger impact in other roads connected to Urip Sumoharjo, Darmo, and Panglima Sudirman.

The travel time is one indicator of traffic congestion.[3] Travel time efficiency is a comparison of the actual travel time with the expected travel time.[4] Traffic congestion affects the travel time delay so that it becomes longer. Based on the above issue, it is necessary to reduce congestion to improve travel time efficiency. Reducing the amount of travel time enables transport users to spend their saved time more productively and enjoyably.

The Transportation Department of Surabaya City has proposed various solutions to reduce traffic congestion, such as the plan to develop mass public transportation, including a tram named Surotram. Unfortunately, the Surotram project is still having difficulty because of the lack of budget[5] and still has not been realized. The Transportation Department of Surabaya City has also developed another public transportation, Suroboyo Bus. Suroboyo Bus gains much interest from Surabaya citizens because it is using plastic waste such as used drinking bottles or glasses of mineral water as payment. This payment method is a form of collaboration with the Department of Sanitation and Public Green Space to reduce and recycle plastic waste.[6] However, the use of the Suroboyo Bus still needs to be increased again to reduce congestion in Surabaya because of its limited travel routes.[7]

System dynamics modeling can influence an organization's performance and help to reflect on potential strategic actions in the future.[8] This research contributes to formulating relationships between several variables related to travel time efficiency, modeling the dynamic behaviour of travel time and traffic congestion, and building scenario models to improve travel time efficiency and reduce traffic congestion in Surabaya, Indonesia. Some references related to travel time and traffic congestion were used as basic knowledge in developing the model. Model validation was carried out to check the basic model's validity with historical data. With the validated model, several potential strategies to increase urban mobility and reduce traffic congestion were tested and evaluated through structural scenarios. The scenario development includes bus rapid transit (BRT) and tram project that initially the plan of the Transportation Department of Surabaya City.[9,10]

2 LITERATURE REVIEW

a System Dynamics Simulation to Improve Travel Time Efficiency

Traffic congestion is a condition when traffic flow exceeds road capacity, resulting in vehicle queues and a decrease in speed.[11] Surabaya experienced severe traffic congestion and was ranked fourth in the Stop-Start level released by Castrol Magnetec.[12] Travel time is related to congestion, which is travel time and delay that exceeds normal or free flow conditions.[13] Traffic performance is measured by the degree of saturation. The increase in travel time is influenced by the high degree of saturation and delay due to increased transportation activities.[14] Degree of saturation is a comparison between traffic volume and road capacity. Traffic

Dynamic Simulation Model: Travel Time Efficiency

volume is the accumulation of the average daily vehicle volume per hour multiplied by a conversion factor that is passenger car equivalent (PCE). PCE is the value of equality of other vehicles to vehicle passenger cars or light vehicles, while other vehicles are motorcycles and heavy vehicles including all medium buses to big buses and all types of trucks. Table 12.1 shows the value for the urban divided road PCE based on Indonesia Highway Capacity Manual (IHCM). Delay performance is obtained from the comparison between the standard maximum delay and the travel time delay.

Travel time was calculated using IHCM Volume Delay Formula (VDF),[15] which proved to be more accurate than the United States Bureau of Public Roads (BPR) VDF.[16] IHCM VDF can be expressed by an equation as follows:

$$T = T_0 + \alpha_1 \times (Q/C)^\beta + \alpha_2 \times (Q/C)$$

where
  T: travel time (minutes)
  T_0: travel time in free flow (minutes)
  Q: traffic volume (passenger car unit/hour)
  C: road capacity (passenger car unit/hour)
  α_1, α_2, β: parameters which are obtained by using the least squares method as shown in Table 12.2.

TABLE 12.1
Passenger car equivalent for urban divided road[17]

Road Type	PCE of Heavy Vehicle	PCE of Motorcycle
2 lanes, 1 way (2/1)	1.3	0.40
4 lanes, 2 way (4/2 D)	1.2	0.25
3 lanes, 1 way (3/1)	1.3	0.40
6 lanes, 2 way (6/2 D)	1.2	0.25

TABLE 12.2
Parameters of IHCM VDF for multilane roads[18]

Speed in Free Flow	α_1	α_2	β
40 km/hour	0.24	0.84	8.04
50 km/hour	0.21	0.70	8.20
60 km/hour	0.18	0.61	8.08
70 km/hour	0.16	0.54	7.84
80 km/hour	0.14	0.48	7.32

System dynamics is a model development approach in studying a system's complexity and dynamic behavior.[19] The structure of a system dynamics modeling is represented by a causal loop diagram.[20] The model created can be tested based on the impact of the strategies and conditions applied to the system.[21] The system dynamics model reflects changes through causal relationships between variables by positive or negative impacts. The model can be tested through simulations; then the outcomes are known by changing or modifying various existing parameters and variables.[22] Simulations also have several advantages over mathematical models because they can be experimented with without any real risk to the system and used for studies due to the wide variety of inputs.

b Scenario Planning to Reduce Traffic Congestion

System dynamics enables to identify and define traffic problems. Model validation is required to ensure that the model represents the actual system. After the model is valid, scenario development is conducted based on the proposed strategy. Scenario development enables researchers to test the assumption of the model from a future perspective[23] and refers to indicators that affect travel time efficiency and congestion levels. The scenario includes a bus rapid transit (BRT) project, a tram project with highway expansion, and a tram project without highway expansion.

3 MODEL DEVELOPMENT

Casual loop diagram (CLD) describes the relationship between variables that exist in the system.[24] CLD has the main function to display the causal hypothesis of each variable that has been found so that it can be represented in a better structure.[25] CLD in Figure 12.1 has three reinforcing loops (R1, R2, and R3) and two balancing

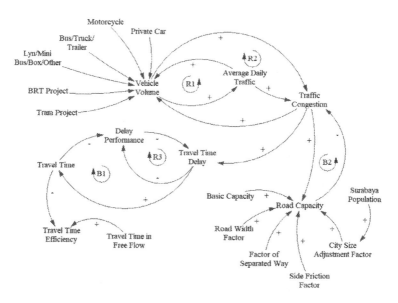

FIGURE 12.1 CLD of travel time efficiency and traffic congestion.

Dynamic Simulation Model: Travel Time Efficiency 157

loops (B1 and B2). An increase in vehicle volume results in an increase in average daily traffic (R1). An increase in average daily traffic results in an increase in traffic congestion (R2). An increase in travel time delay results in a decrease in delay performance (R3). An increase in delay performance results in a decrease in travel time delay, an increase in travel time delay results in an increase in travel time, and an increase in travel time can reduce delay performance (B1). An increase in traffic congestion may result in an increase in the need for additional road capacity (B2).

Vehicle volume is the accumulation of the vehicle variable: Motorcycle, Private Car, Lyn/Minibus/Box/Other, and Bus/Truck/Trailer. Motorcycle and Private Car variables each are the number of motorcycles and the number of private cars. Lyn/Minibus/Box/Other variables are non-private vehicles that are included in light vehicles such as lyn, minibuses, mini trucks, boxcars, and non-motorized vehicles. The Bus/Truck/Trailer variable shows non-private heavy vehicles, such as large buses, trucks, and trailers. Average Daily Traffic variable will be divided by hours per day as Traffic per Hour. Surabaya population is required to determine the value of city size adjustment factor for road capacity.

4 MODEL VALIDATION

Model validation requires historical data covering the time horizon of the base model simulation (2000–2020). The data is obtained from the Central Bureau of Statistics, the Department of Transportation Service and other agencies related to the transportation system in Surabaya. Model is validated by checking the error rate and the error variance as shown in the equations as follows:

Error Rate = |Average of simulation data–Average of historical data|/ Average of historical data
Error Variance = |Standard deviation of simulation data–Standard deviation of historical data|/Standard deviation of historical data

A model is valid if the error rate is less-than or equal to 5% and the error variance is less than or equal to 30%.[26] The error rate of some variables such as traffic per hour and Surabaya population are as follows:

Error Rate of "Traffic per Hour" = |5308.96–5235.28| / 5235.28 = 1.4074%
Error Rate of "Surabaya Population" = |2903723.33–2925994.37| / 2925994.37 = 0.7611%

The error variance of traffic per hour and Surabaya population are as follows:

Error Variance of "Traffic per Hour" = |1404.65–1354.78| / 1354.78 = 3.6809 %
Error Variance of "Surabaya Population" = |176906.19–175844.72| / 175844.72 = 0.6036%

The comparison between the simulation results and the historical data of traffic per hour and Surabaya population can be seen in Figures 12.2 and 12.3.

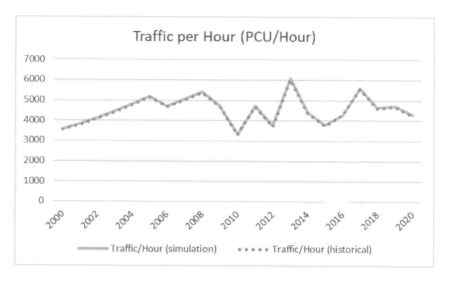

FIGURE 12.2 Comparison of simulation result and historical data of traffic per hour.

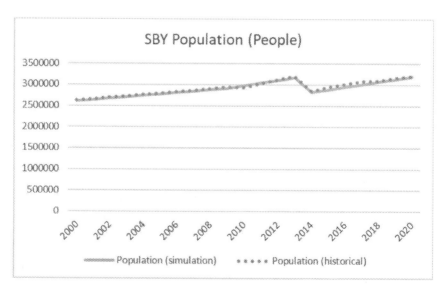

FIGURE 12.3 Comparison of simulation result and historical data of Surabaya population.

5 SCENARIO DEVELOPMENT

This section presents several scenarios conducted to reduce traffic congestion and improve travel time efficiency. Some strategies are carried out to reduce traffic congestion and thus can improve travel time efficiency. The scenarios are developed by modifying the model structure and parameters to shows the impact of various strategies.

Dynamic Simulation Model: Travel Time Efficiency

5.1 Bus Rapid Transit Development Scenario

Bus rapid transit (BRT) is a high-quality bus system that is fast, convenient, and cost-effective.[27] The scenario of BRT was developed based on the Surabaya government's plan to reduce congestion. It had been implemented in Surabaya from September until December 2016.[28] By the implementation of BRT, the percentage of private vehicle users that will switch to BRT is around 7.5%. The percentage of BRT users is projected to grow 2% annually by the changes in user behavior and the number of additional bus fleets. The simulation result of BRT scenario towards traffic congestion and travel time efficiency is shown in Figures 12.4 and 12.5.

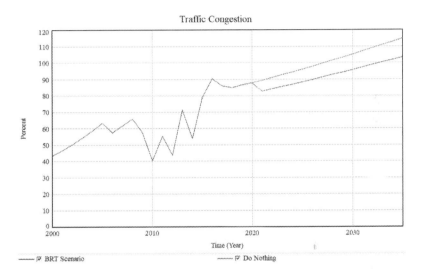

FIGURE 12.4 Comparison of simulation result of traffic congestion for BRT scenario.

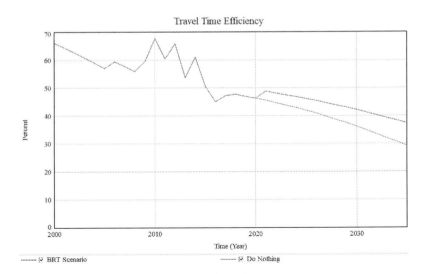

FIGURE 12.5 Comparison of simulation result of travel time efficiency for BRT scenario.

5.2 TRAM DEVELOPMENT WITHOUT HIGHWAY EXPANSION SCENARIO

The tram development scenario was designed to reduce congestion. This scenario was one of the plans of Surabaya government entitled Surabaya Mass Rapid Transportation (SMART).[29] The initial percentage of private vehicle users that will switch to tram is 19%. The tram users will grow around 0.19% annually. The original plan of the tram project will be using one lane of the road for each way. Therefore, the road capacity will decrease with the reallocation of the road space for tram project. Previous study also predicted that this circumstance would increase the traffic congestion.[30] The simulation result of tram without highway expansion scenario towards traffic congestion and travel time efficiency is shown in Figures 12.6 and 12.7.

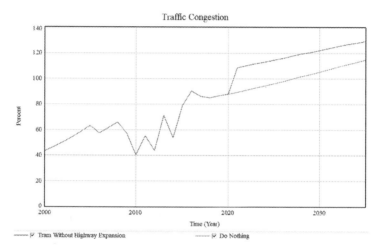

FIGURE 12.6 Comparison of simulation result of traffic congestion for tram without highway expansion scenario.

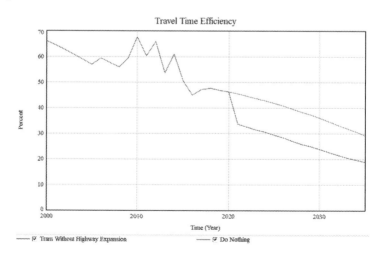

FIGURE 12.7 Comparison of simulation result of travel time efficiency for tram without highway expansion scenario.

Dynamic Simulation Model: Travel Time Efficiency

5.3 Tram Development with Highway Expansion Scenario

This scenario is similar to the previous scenario, but it will be developed with the addition of lanes to support the tram project. In a previous study, the best practice for road capacity expansion is to prioritize new transit projects over addition to regional roadway networks.[31] The road capacity for other transport modes will not be reduced. The result shows that it can decrease the traffic congestion better than the BRT project. The simulation result of tram with highway expansion scenario towards traffic congestion and travel time efficiency is shown in Figures 12.8 and 12.9.

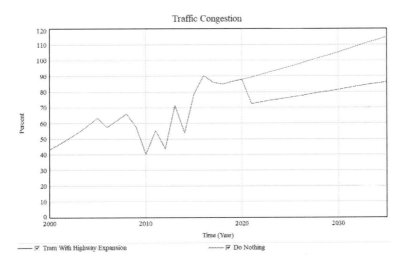

FIGURE 12.8 Comparison of simulation result of traffic congestion for tram with highway expansion scenario.

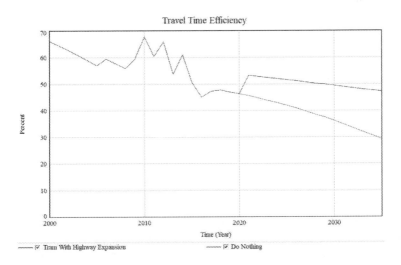

FIGURE 12.9 Comparison of simulation result of travel time efficiency for tram with highway expansion scenario.

6 CONCLUSION AND FURTHER RESEARCH

This research is developed to model the Surabaya transport system to reduce traffic congestion and increase travel time efficiency. The model development is carried out based on the current condition of the actual system. Based on the developed model and scenarios, the number of private vehicles brings a significant influence on traffic congestion. The reduction in road capacity also could increase traffic congestion because of the current condition of the traffic volume. Several strategies can be implemented to reduce congestion such as BRT and tram development. BRT is predicted to reduce the number of private vehicle users by 7.5% and will grow 2% annually after the first year. The original plan of tram development in Surabaya will reduce the current road capacity for tramways. While the tram project was proposed to reduce the traffic volume, but the traffic congestion is increased because of the reduction in road capacity. If the tram development comes with highway expansion, the current road capacity will not be decreased, and the traffic congestion can be better than the BRT scenario. Further research is required to develop a comprehensive transport system with the consideration of other factors such as investment, infrastructure, or accessibility.

NOTES

1. Surabaya City Government. *Regulation of Surabaya Mayor No. 27 Year 2018 Regarding Local Government Work Plan (RKPD) Surabaya City Year 2019*; 2019.
2. Priyambodo. Correlation Analytic of Vehicles and GDP on East Java Province (Analisis Korelasi Jumlah Kendaraan dan Pengaruhnya Terhadap PDRB di Provinsi Jawa Timur). *War Penelit Perhub.* 2018;30:59–65. doi:10.25104/warlit.v30i1.634.
3. Susilo BH, Imanuel I. Traffic congestion analysis using travel time ratio and degree of saturation on road sections in Palembang, Bandung, Yogyakarta, and Surakarta. *MATEC Web Conf.* 2018;181. doi:10.1051/matecconf/201818106010.
4. Grison E, Gyselinck V, Burkhardt JM. Exploring factors related to users' experience of public transport route choice: Influence of context and users profiles. *Cogn Technol Work.* 2016;18(2):287–301. doi:10.1007/s10111-015-0359-6.
5. Puspita R. The Tram Payment Scheme in Surabaya Is Uncertain (Skema Pembayaran Trem di Surabaya Belum Jelas). *Republika.co.id.* https://republika.co.id/berita/owavp0428/skema-pembayaran-trem-di-surabaya-belum-jelas. Published September 15, 2017. Accessed January 17, 2021.
6. Utomo N, Sholichin I, Wiradiansyah P. Suroboyo bus fares sufficiency with willingness to pay aspects (Case Study: Purabaya Bus Station-Rajawali Shelter). *J Civ Eng Sci Technol.* 2020;1(1):16–26. http://ci-tech.upnjatim.ac.id/index.php/ci-tech/article/view/12.
7. Kurniawan AA, Prabawati I. Implementation of Suroboyo bus at the Surabaya City transportation agency (Implementasi Suroboyo Bus di Dinas Perhubungan Kota Surabaya). *J Mhs Unesa.* 2018;6(9).
8. Torres JP, Kunc M, O'Brien F. Supporting strategy using system dynamics. *Eur J Oper Res.* 2017;260(3):1081–1094. doi:10.1016/j.ejor.2017.01.018
9. Faiq N. Surabaya Citizen Long for the Affordable and Comfortable Bus Rapid Transit (Warga Surabaya Rindukan Bus Rapid Transit yang Murah dan Nyaman). *Surya.* https://surabaya.tribunnews.com/2016/12/11/warga-surabaya-rindukan-bus-rapid-transit-yang-murah-dan-nyaman?page=all. Published December 11, 2016.

10. Imani PN. Measuring Readiness and Willingness to Pay (WTP) of Surabaya Mass Rapid Transit (SMART), Monorail and Tram: a Survey. *J Tek ITS*. 2015;4(1).
11. Cambridge Systematics Inc. *Traffic Congestion and Reliability: Trends and Advanced Strategies for Traffic Congestion Mitigation*; 2005. doi:10.1080/10915810500434209.
12. Zainuddin. Surabaya Ranks Fourth on World's Most Congested City (Kemacetan Surabaya Masuk Empat Besar Dunia). *Surya*. https://surabaya.tribunnews.com/2015/02/06/kemacetan-surabaya-masuk-empat-besar-dunia. Published February 6, 2015.
13. Lomax TJ, Turner S, Board G, et al. *Quantifying Congestion, Volume 1: Final Report*. Washington, DC; 1997. https://trid.trb.org/view.aspx?id=475257.
14. Susilo BH, Imanuel I. Traffic congestion analysis using travel time ratio and degree of saturation on road sections in Palembang, Bandung, Yogyakarta, and Surakarta. *MATEC Web Conf.* 2018;181. doi:10.1051/matecconf/201818106010.
15. Directorat General of Bina Marga. Indonesian Highway Capacity Manual (IHCM). 1997.
16. Irawan MZ. Implementation of the 1997 Indonesian Highway Capacity Manual (MKJI) Volume Delay Function. *J East Asia Soc Transp Stud*. 2010;8:350–360. doi:10.11175/easts.8.350
17. Directorat General of Bina Marga. Indonesian Highway Capacity Manual (IHCM). 1997.
18. Irawan MZ. Implementation of the 1997 Indonesian highway capacity manual (MKJI) volume delay function. *J East Asia Soc Transp Stud*. 2010;8:350–360. doi:10.11175/easts.8.350
19. Sterman J. *Business Dynamics, Systems Thinking and Modeling for a Complex World*. New York: McGraw-Hill, 2000. doi:10.1016/s0016-0032(99)90164-3.
20. Das D, Dutta P. A system dynamics framework for integrated reverse supply chain with three way recovery and product exchange policy. *Comput Ind Eng*. 2013;66(4):720–733. doi:10.1016/j.cie.2013.09.016.
21. Wei S, Yang H, Song J, Abbaspour KC, Xu Z. System dynamics simulation model for assessing socio-economic impacts of different levels of environmental flow allocation in the Weihe River Basin. China. *Eur J Oper Res*. 2012;221(1):248–262. doi:10.1016/j.ejor.2012.03.014.
22. Sugiarto F, Buliali JL. Implementation of system simulation for production process optimization in fish canning company (Implementasi Simulasi Sistem untuk Optimasi Proses Produksi pada Perusahaan Pengalengan Ikan). *J Tek ITS*. 2012;1:236–241.
23. Forrest J. System dynamics, alternative futures, and scenarios. In: *The 16th International Conference of The System Dynamics Society*, Quebec, Canada; 1998.
24. Kiani B, Gholamian MR, Hamzehei A, Hosseini SH. Using causal loop diagram to achieve a better understanding of E-business models. *Int J Electron Bus Manag*. 2009;7(3):159–167. http://ijebm.ie.nthu.edu.tw/IJEBM_Web/IJEBM_static/Paper-V7_N3/A02.pdf.
25. Sterman J. *Business Dynamics, Systems Thinking and Modeling for a Complex World*. New York; McGraw-Hill, 2000. doi:10.1016/s0016-0032(99)90164-3.
26. Barlas Y. Formal aspects of model validity and validation in system dynamics. *Syst Dyn Rev*. 1996;12(3):183–210. doi:10.1002/(sici)1099-1727(199623)12:3<183::aid-sdr103>3.3.co;2-w.
27. Institute for Transportation and Development Policy. What is BRT? www.itdp.org/library/standards-and-guides/the-bus-rapid-transit-standard/what-is-brt/. Accessed 2016.
28. Faiq N. Surabaya Citizen Long for the Affordable and Comfortable Bus Rapid Transit (Warga Surabaya Rindukan Bus Rapid Transit yang Murah dan Nyaman). *Surya*. https://surabaya.tribunnews.com/2016/12/11/warga-surabaya-rindukan-bus-rapid-transit-yang-murah-dan-nyaman?page=all. Published December 11, 2016.

29 Imani PN. Measuring readiness and willingness to pay (WTP) of Surabaya mass rapid transit (SMART), monorail and tram: a Survey. *J Tek ITS*. 2015;4(1).
30 Muhis Z, Herijanto W. Traffic management regarding trams on Darmo Street Surabaya (Manajemen Lalu Lintas Akibat Trem Di Jalan Raya Darmo Surabaya). *J Tek ITS*. 2014;3(1):E37–E42.
31 Ewing R, Proffitt D. Improving decision making for transportation capacity expansion: Qualitative analysis of best practices for regional transportation plans. *Transp Res Rec*. 2016;2568(November):1–8. doi:10.3141/2568-01.

Index

A

Abstractive summarization, 108
Accessible and important precious media, 121
Accuracy, 129
Accurate, 121–122
Adaptive mean shift algorithms, 126
Aggression, 45
AGSO, 133
AHP, 140, 141, 142
    Algorithm for entire text extraction process, 128
    Genetic algorithms, 3
Algorithms, 2
Applied statistics, 2
Arbitrary surfer, 22
Artificial intelligence, 2
Artificial neural networks 3, 11

B

Background, 123
Balancing loop, 99
Basic rules and techniques for process mining, 3
Basis pursuit denoising, 127
Bayesian networks, 3
Bayesian probability theory, 59
Behavioral patterns, 9
    Behavioural pattern test, 100
BEP, 37
Bias, 62, 63
Bias variance trade-off, 63
Big data, 4
Binarization method, 125
Binary decisions, 124
Blind direction-finding, 122
BRT, 154, 156, 159, 161–162

C

Calrank, 24
Car, 98, 101, 105
Car detection, 70
Cardiovascular disease, 139, 140, 142
Causal loop diagram (CLD), 99, 156
    Causal loop diagram of Surabaya transportation system, 99
Channel out punctuation, 110
Classification, 8, 85, 58, 66, 67, 73
    Classification strategies, 11
Clustering, 8, 58, 68, 84
    Cluster analysis, 3
COCO database, 72

Computer vision, 70
Congestion, 97, 98, 99, 101, 102, 103, 104, 105
Construct a link, 24
Convergence speed, 26, 32
Credit cards, 47, 48
Curse of dimensionality, 65
Curve fitting, 60

D

Database, 37
    Database theory, 2
Data collection, 109
Data mining, 1
    Data mining schemes, 2
    Data mining strategies, 2
Data preprocessing, 109
Data visualization strategies, 3
DB-SU, 89
DBSCAN clustering, 88
De-blurring, 124
Decision trees, 3
Degraded backgrounds, 125
Degraded images, 133
Detection and further binarization, 121
Detection of anomaly, 2
Diabetic retinopathy, 84, 85, 87
Diet recommendations, 140, 145
Dimensionality, 65
    Data flow design, 75
Distortion, 127
Diversity of fonts and other properties, 122
Documents get the same rank 28
    Feature extraction, 73

E

Edge-detection–based method, 125
Educational data mining, 5
E-learning system, 1
Embedded Zerotree of Wavelet (EZW), 124
Enhancement, 126
Envisioning, 110
Error rate, 154
Error variance, 100, 101, 154
Evaluation, 131
Event data, 1
Event log data, 1
Evolutionary programming, 3
Existing techniques, 129
Extraction process, 128
Extractive summarization, 108

165

F

F1-score (%) of proposed and existing methods, 132
False positive, faster R-CNN, 129
Feature-based connected components, 125
Features of the image, 122
Fetched as per the requirement of computer vision, 121
Frame, 126
Further streamlined via the use of an optimization method, 123
Fuzzy TOPSIS, 144, 145

G

Gabor transform (GT), 122
Getlinks, 24, 25
    GoogleNet, 78
Graphical user interface, 52

H

Hybrid, 84
Hyperlipidemia, 140
Hypermedia, 5
Hypertension, 140
Hypotheses, 7
Hyperlinked, 22, 33

I

Information, 2
Information retrieval system, 19, 20, 21, 33
Inverted index, 37
Iris data, 64, 66, 68
Isolated hubs are eliminated, 25
Iterations, 23, 24, 26, 27

K

Keras API, 75
K-means clustering, 87
K-nearest neighbor, 50
Knowledge, 7, 85

L

Latent semantic analysis, 112
Lemmatizer, 40
Lexical likeness, 112
Light rail transit (LRT), 99, 101, 102, 104, 105
    LRT scenario on the use of public transportation, 103
    LRT scenario on traffic congestion, 103
Linear discriminant analysis, 65
Linear regression, 3
Link structure, 20, 24, 25, 32
Logistic regression, 48
Lower assessment score, 115
Luminance, 125

M

Machine learning, 2, 47, 48
Mass rapid transit, 99, 101, 104, 105
    MRT scenario on the use of public transportation, 102
    MRT scenario on traffic congestion, 102
Mean comparison, 100, 101
Methods, 130
MobileNet, 82
Model, 98, 99, 100, 101, 104
Model development, 156, 162
Model validation, 154, 156–157
Modern information system, 1
Moister factor, 20, 22, 23, 24, 26, 27, 28, 29, 30, 31, 32
    Object acquisition, 71
    Object detection, 70
Monitoring of learning progress, 7
    Resource allocation planning, 7
Morphological component analysis, 127
Motorcycle, 98, 101
Multi-disciplinary approach, 2

N

Nearest neighbor, 3
Neural networks, 122
Non-text part, 129
Normalization, 40, 110

O

Optic, 85
Optimal, 142
    Optimal search results, 21
Outbound links, 20

P

Page Rank, 112
Parameter scenario, 99
Passenger, 98, 99, 101, 102, 103, 105
Passenger car equivalent, 155
Pattern, 57
Pattern recognition, 2, 57
Polynomial, 61
Population, 100
Precision, 44
Predictive analytics, 2
Principle component analysis, 65
Private transportation, 98, 99

Index

Probability theory, 58
Process mining, 1
Proximity, 44
Public transportation, 98, 99, 100, 101, 102, 103, 104, 105

Q

Q-matrix, 10
Quantitative measuring parameters, 7
Query independent ranking, 19, 20, 21

R

Random forests, 49
Random surfer model, 24
Ranking, 42
Recall, 44
Recorded data, 1
Recursive algorithm, 20
    R-CNN, 76
        Faster R-CNN, 71, 76
        Region proposal networks, 77
Regression analysis, 3
Reinforcing loop, 99
Repositories, 2
Retrieval, 126
Root mean squared error, 62
    Region of interest, 71
ROUGE-N metric, 115

S

Sack counting system, 70–75
Search engine, 36
Semantically like expressions, 115
Sequence diagram, 75
SERPs, 36
Simulate the behavior of the hyperlink, 24
Simulation, 99, 100
Specific worth crumbling, 113
Stemming, 37
Structural scenario, 99, 101
Submodel, 104
Supervised learning, 48
Support vector machine, 3
Surabaya, 97, 98, 104
    Surabaya City Transportation Department, 98

Suroboyo bus, 98, 100, 101
System dynamics, 98, 154, 156

T

TensorFlow, 75
    TensorFlow 2.0, 76
Term-archive network, 112
Term frequency inverse document frequency, 110
Text mining, 8
Text rank algorithm, 112
Text summarization, 108
TF-IDF methodology, 112
Toggle state, 24, 28
Tokenization, 39, 109
TPU silicon, 76
Tram, 154, 156, 160–162
Transit-oriented development, 98, 99, 101
Travel time, 153–162
Travel time efficiency, 153–154, 156, 158–162

U

Unigrams, 115, 118
Unmistakable procedures, 118
Utilization, 7
    Utilization of association schemes, 2

V

Value of prediction, 8
Variance, 62, 63
Vehicle, 97, 98, 99, 104, 105

W

Web graph, 25
Web structure-based ranking, 20
Weighted naïve bayes algorithms (WNBCs), 123
WordNet, 42

Y

YOLO model, 71

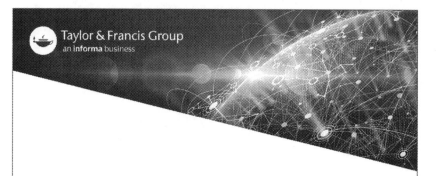

Taylor & Francis eBooks

www.taylorfrancis.com

A single destination for eBooks from Taylor & Francis with increased functionality and an improved user experience to meet the needs of our customers.

90,000+ eBooks of award-winning academic content in Humanities, Social Science, Science, Technology, Engineering, and Medical written by a global network of editors and authors.

TAYLOR & FRANCIS EBOOKS OFFERS:

- A streamlined experience for our library customers
- A single point of discovery for all of our eBook content
- Improved search and discovery of content at both book and chapter level

REQUEST A FREE TRIAL
support@taylorfrancis.com

Printed in the United States
by Baker & Taylor Publisher Services